Micro Quick!

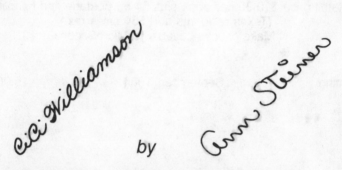

by

CiCi Williamson and Ann Steiner
NATIONWIDE NEWSPAPER COLUMNISTS

FEATURING:

- Quick and easy microwave recipes
- Convenience food information and charts
- Special Sections:
    Baby Food
    Herb and Flower Drying
    Microwave to Grill
    Homemade Ice Creams

314 recipes *"Quick"-er* than 20 minutes plus 28 of our favorites which take a little longer.

Illustrations by CiCi Williamson
Cover Illustration by Melanie J. English

To order additional copies of **Micro Quick,** or our basic **Microwave Know-How** book (formerly **MicroScope Savoir Faire**), write to:

MicroScope
P.O. Box 79762
Houston, Texas 77279

Retail price $10.95 per book plus $1.50 postage and handling;
(Texas residents add 55¢ sales tax.)
Make checks payable to MicroScope.

1st Printing   September, 1984   15,000 copies

ISBN #0-9607740-1-7
Library of Congress Card Number 84-061145

Copyright ©1984 by
CiCi Williamson and Ann Steiner

Printed in U.S.A.
Hart Graphics
8000 Shoal Creek Blvd.
Austin, Texas 78758

All Rights Reserved

This book or any portion of this book may not be reproduced in any form without written permission of the authors, except for use in newspaper or magazine reviews, for which credit must be given.

THIS BOOK IS DEDICATED TO OUR READERS

We swore we'd never write another book! But readers of **MicroScope,** our microwave newspaper column, have been requesting that we write a quick and easy version to complement our first **Microwave Know-How** book, **MicroScope Savoir Faire**. Many people have told us that they cut out our column every week, and that their files are bulging! We've included many of the favorite recipes from our weekly **MicroScope** columns.

As home economists and microwave specialists, we saw a need for good reliable information in small doses. In 1979 we came up with the idea of writing a microwave cooking column for our "hometown" newspaper, the *HOUSTON CHRONICLE*. Food editor Ann Criswell liked our idea and it became a reality. Today our column can be read coast to coast by millions. We've discovered that the "American Dream" can still come true!

A SPECIAL THANKS TO EACH OTHER

Before we wrote our first book, we checked a writer's manual out of the library which advised, "Never co-author anything." Luckily you can't believe everything you read! Through it all, we're still best friends.

Working together, we're on the same wave length. One of us starts a sentence and the other can finish it. When we teach microwave cooking classes together, one wordlessly puts a needed spatula into the other's outstretched hand.

Every woman needs a wife!

FOREWORD

Because our newspaper column, **MicroScope,** appears in so many states, we offered our food editors the opportunity to have their comments included in **Micro Quick.** Here is what some of them said.

"Ann, CiCi and I have come a long way together since they first wandered into my office five years ago. After adopting the helpful techniques in **MicroScope**—such as cooking rice and sautéing ground beef in a colander—we have been ready for more. **Micro Quick** is it!"

Ann Criswell, Food Editor
HOUSTON CHRONICLE

"Ann and CiCi are friends, as well as authors, who help out in East Tennessee kitchens each week!"

Louise Durman, Food Editor
THE KNOXVILLE NEWS-SENTINEL

"Readers of our home pages tell me the good-sounding recipes in the **MicroScope** column are reason enough to want to own a microwave oven! I'm sure everyone who has found **MicroScope Savoir Faire** a great help will want the new collection of recipes, too."

Anne Thomas, Food Editor
HUNTERDON COUNTY (NJ) DEMOCRAT

"Ann and CiCi, through their column, **MicroScope,** offer not only microwave recipes, but information on the how's and why's of microwaving. This enables the readers to understand the reasoning behind the recipes and extend that information into their own cooking styles and favorite family recipes. The column is a useful tool for any microwave owner."

Sheila Friedeck, Food Editor
THE BEAUMONT ENTERPRISE

"It's impossible to imagine how both beginners and experienced cooks could have ever mastered the intricacies of microwave cooking without **MicroScope Savoir Faire,** since all of us are now living in the 'fast lane.'"

Jeanne Lively, Food Editor
LUBBOCK AVALANCHE JOURNAL

"**MicroScope** is so easy to use. It is the readers' favorite column in our food section."

Louis Mahoney, Food Editor
RICHMOND NEWS-LEADER

"CiCi and Ann succeed in taking the mystery out of microwaving by adding the magic of their imagination and meticulous testing of recipes. **MicroScope** recipes are for those who want to use and enjoy their microwave ovens every day, whether they're in a hurry or not."

Jo Ann Vachule, Food Editor
FORT WORTH STAR-TELEGRAM

"**MicroScope** is the microwave column I've been searching for. It's great and the JOURNAL readers love it!"

Jane Mengenhauser, Food Editor
THE JOURNAL NEWSPAPERS (VA & MD)

"Your microwave cooking column has become a popular feature in the STATESMAN-JOURNAL's weekly food section. Its subject variety and practical advice make it a useful tool for the microwave cook."

Mary J. Parkinson, Food Editor
(SALEM) STATESMAN-JOURNAL

"Ann and CiCi have done a great deal to help the microwave oven change its image from a convenient 'reheater' to a useful cooking tool."

Rhonda Hoeckley
THE EAGLE (BRYAN-COLLEGE STATION)

"The **MicroScope** column by Ann Steiner and CiCi Williamson has been a regular feature in THE SACRAMENTO BEE food pages since 1981. Reader response has been excellent to this practical approach to microwave cooking."

Dorothy Sorenson, Food Editor
THE SACRAMENTO BEE

"So many Florida residents have microwave ovens, and **MicroScope** has really taught us how to use them!"

Ann McDuffie, Food Editor
THE TAMPA TRIBUNE

"We look forward to your new book. The **MicroScope** column has been a good addition to our weekly food section. Many readers have expressed how much they enjoy it."

Chris Wheeler, Food Editor
HAMILTON JOURNAL NEWS (OHIO)

ABOUT THE AUTHORS

Ann Steiner and CiCi Williamson are co-authors of **MicroScope,** a microwave cooking column featured weekly in newspapers across the nation. Besides *Micro Quick,* they have also written a *Microwave Know-How* book, *MicroScope Savoir Faire.*

As home economists and microwave specialists, they have conducted cooking schools for major microwave oven manufacturers, department stores and gourmet cookware shops in various Texas cities as well as other states throughout the U.S. In addition, they have appeared on local TV programs as well as nationwide cable television sharing their microwave expertise.

Their professional memberships include the International Microwave Power Institute, American Home Economics Association and Home Economists in Business. The authors are included in *Notable Women of Texas.*

Ann received a bachelor's degree from Miami University, Oxford, Ohio, and her master's degree from Ohio State University. CiCi received her degree from the University of Maryland. A native of Dallas, Texas, she is Feature Editor of The Key magazine of Kappa Kappa Gamma.

Ann and CiCi are from Houston, Texas.

ABOUT THE BOOK

RECIPES

All recipes were developed and tested by us in 650 to 700 watt microwave ovens. All were tasted and rated by our families and friends (only the best ones made it to the book!) In some cases, brand names are used in our recipes. These are our personal preferences and we receive no payment for their mention.

THE SYMBOL "MW"

The word "microwave" can have several meanings—an appliance, a radio wave or a cooking procedure. Throughout *Micro Quick,* we have used the symbol **"MW"** in place of the word microwave when used as a verb meaning "to cook."

STACON COVER

Our Stacon book cover is a washable and unbreakable plastic. Do not leave it in an area of extreme heat (such as in a locked car sitting in the sun).

TABLE OF CONTENTS

Power Levels 8
Determining Microwave Time by Weight 10
The Great Cover Up 11
Frozen Convenience Foods 12
Casserole Technique 15
Reheating Foods 16
APPETIZERS 18
BEVERAGES 29
SOUPS 34
SALADS 41
EGGS & QUICHE 50
PASTA & CHEESE MAIN DISHES 58
MEAT 70
POULTRY 107
FISH & SHELLFISH 115
VEGETABLES 124
RICE & DRESSINGS 153
DESSERTS 158
ICE CREAM 170
QUICK BREADS 174
BAR COOKIES 180
CAKES 184
PIES 193
CANDY & GIFTS 204
BABY FOOD 211
HERB & FLOWER DRYING 214
INDEX 217

POWER LEVELS

To cook food properly in a microwave oven, you must use the best power level for that type of food. If you do not understand the power levels on your oven, you may not be setting it correctly.

The International Microwave Power Institute, of which we are members, has established five universal power settings. **These are the ones we use for our recipes:**

> 100% - High
> 70% - Medium-high
> 50% - Medium
> 30% - Medium-low
> 10% - Low

Before using the recipes in this book, newspapers, magazines, or other microwave cookbooks, be sure you understand which power settings on your oven correspond to these.

If your microwave oven is one with "touch controls" which has numbers 1 to 10, your power settings are easy to understand. Just add a zero to each number, and you will see the percentages of power. Use number 7 for 70%, 5 for 50%, 3 for 30% and 1 for 10%.

If your microwave oven has controls displayed by words, look at our simplified chart on page 9 to determine your percentages of power for the universal power settings.

We have not listed 100% power on our chart, because all ovens have this setting. **"High, Full Power and Normal" all mean 100% power.** If you turn your oven on without selecting a lower power level, it will always microwave at 100% power.

BEST POWER LEVELS FOR FOODS

HIGH POWER (100%): Ground meats; chicken and turkey; fish and shellfish; bacon; fruits and vegetables; water-based soups or casseroles; pasta; pie crust; reheating bread products; and drying herbs and flowers (see pages 214 to 216).

MEDIUM-HIGH POWER (70%): Rib roast; duck and Cornish game hens; quiche and foods containing cheese, cream-based soups or casseroles; brownies; cheesecake; reheating leftovers; and microwaving frozen convenience foods.

MEDIUM (50%): Ham and pork; stew; eye round roast; eggs and custards; softening cream cheese; and melting chocolate.

MEDIUM-LOW (30%): Less tender beef pot roasts; lamb; simmering chili, stew and spaghetti sauce; rice and cereal; and defrosting.

LOW (10%): Defrosting large roasts or turkeys; keeping food warm; and proofing yeast dough.

POWER LEVEL SETTING CHART

We have not listed 100% power on our chart, because all ovens have this setting. "High, Full Power and Normal" all equal 100%.

MANUFACTURER	10% POWER	30% POWER	50% POWER	70% POWER
Amana	Warm, Low	Simmer, Defrost	Slo Cook	Roast, Bake, Medium High
Caloric	Warm, Low	Simmer, Slo-Cook, Defrost	Medium	Roast, Bake
Frigidaire	Warm	Simmer, Defrost	Medium	Medium High
GE/Hotpoint	Warm	Low, Defrost	Medium	Medium High
JennAir	Warm	Low, Medium-Low, Simmer	Medium, Defrost	Medium High, Bake
Kelvinator	Low	Defrost, Medium-Low	Medium	Medium High
Magic Chef	Low	Defrost	Medium Low, Defrost	Medium High
Montgomery Ward	Keep Warm	Low, Defrost	Simmer, Stew	Roast, Bake
Norelco	Warm	Simmer, Defrost	Simmer, Defrost	Bake, Roast
O'Keefe & Merritt	Warm	Simmer, Low-Defrost	Simmer, Medium	Medium High, Roast
Panasonic, Quasar, J. C. Penney's	Warm	Low, Defrost	Medium Low	Medium
Roper	Warm	Low, Simmer, Defrost	Defrost, Simmer	Roast
Sanyo	Warm	Defrost	Simmer, Braise, Defrost	Roast
Sears/Kenmore	Warm, Low	Serve, Defrost	Simmer, Defrost	Medium, Roast, Bake
Sharp	Warm, Low	Simmer, Low-Defrost	Medium, High Defrost	Medium High, Roast
Sunbeam	Keep Warm	Simmer	Defrost	Medium High, Roast, Bake
Tappan	Keep Warm, Low	Simmer, Low-Defrost	Medium, High Defrost	Medium High
Thermador/Waste King	Warm	Low, Simmer	Slo Cook, Low	Roast, Medium
Toshiba	#1	#3	#5	#6 or 7
Whirlpool	Low	Defrost, Medium-Low	Medium	Medium High
White Westinghouse	Warm	Low, Defrost, Simmer	Medium Low, Slo-Cook, High Defrost	Medium

DETERMINING MICROWAVE TIME BY WEIGHT

An important principle of microwave cooking is that **the more food you put into a microwave oven, the longer it takes to cook.** If you weigh food on a kitchen scale, you can determine the microwave time for cooking it perfectly.

If you have not been using a scale, you can only guess at the results. Potatoes, for example, may be super spuds one night and disaster duds another!

Different foods require differing amounts of cooking time and also various power levels. Use the following charts for microwaving meats and vegetables.

MEAT MICROWAVING CHART

MEAT	POWER LEVEL	MINUTES PER POUND
Ground beef, lamb	100%	5 to 6
Ground pork, sausage	100%	6 to 7
Tender beef roasts:		
Eye of round	50%	8 to 9
Tenderloin	50%	9 to 11
Standing rib	50%	11 to 13
Rolled rib	50%	13 to 15
Sirloin tip	50%	15 to 17
Chuck roast, stew meat	30%	20
Pork tenderloin	70%	8 to 10
Pork chops	50%	16 to 18
Ham, precooked	50%	10
Ham, raw	50%	16 to 18
Chicken, turkey	100%	6 to 7
Cornish hens	70%	8 to 9
Fish	100%	3 to 4

FRESH VEGETABLE MICROWAVING CHART

VEGETABLE	POWER LEVEL	MINUTES PER POUND
Corn on the cob	100%	2½ to 3 per ear
Spinach, tomato halves	100%	3 to 4
Artichokes	100%	3 per artichoke
Soft-shelled squash	100%	4 to 5
Diced eggplant, okra, peas	100%	5 to 6
Asparagus, broccoli, Brussels sprouts, cauliflower, baked potatoes, hard-shelled squash	100%	6 to 7
Cabbage, carrots, green beans, whole new potatoes	100%	7 to 10
Whole beets	100%	15 to 18

THE GREAT COVER-UP

When you are covering foods to cook in the microwave oven, do you just reach in the drawer and pull out the first thing handy? Beware! The covering makes a difference in the cooking time and the texture of the finished product.

There are four types of covers we use in microwave cooking: **paper toweling; wax paper; glass lids; and plastic wrap.** Sometimes, we purposely do not use a cover at all.

Paper toweling is used to prevent splattering, such as in cooking bacon or crumb-coated chicken. It also absorbs moisture when used to defrost and reheat bread products. Paper toweling holds in a minimum amount of heat and promotes uniform cooking.

Wax paper is used to hold in more heat and moisture. Meats are often covered with a wax paper tent during or after cooking. Although baked products (such as cakes or brownies) are microwaved uncovered, we cover them with wax paper afterwards to help hold in the heat during standing time.

Glass lids hold in more heat and moisture than does wax paper. On browning dishes, we use glass lids to prevent splattering and to hold in heat (for example: frying an egg). Glass lids are also used when microwaving casseroles, soups and spaghetti sauce. They are not airtight and will sometimes rock if there is too much heat and steam within. A lower power level will prevent a glass lid from rocking.

Plastic wrap holds in the maximum amount of heat and moisture. We don't pierce plastic wrap after covering food because we want to hold in all of the moisture and heat possible. Neither of us has ever had plastic wrap "blow a hole" due to steam build up. Allow some expansion room when placing the plastic wrap over food. We don't use plastic wrap to cover meats since they would have a steamed quality. Fat from the meat can melt the plastic wrap. Fresh vegetables cook best when covered with plastic wrap, and little, if any, water is needed for most vegetables, other than that which clings to the vegetable from washing.

The general rule of thumb is to **"cover items in the microwave that you would cover if cooking conventionally."**

- Baked products such as cakes, pies and bar cookies aren't covered since they require a dry surface.
- Reheating is done under cover!
- When thickening sauces, do not cover.
- Soups and large quantities of beverages will heat more quickly if covered.
- Fish, ground meats and tender cuts of meat are covered with wax paper.
- Plastic wrap is used for fresh veggies and eggs to hold in the maximum amount of heat and moisture, and to promote uniform cooking.
- Oven cooking bags are used for less-tender cuts of meat, since they hold in a maximum amount of heat and moisture. They are more durable than plastic wrap, and will not split or melt. We don't pierce the bag. Leave an opening the size of a quarter when tying the end of the bag. Zip-closing freezer bags can also be used for cooking or reheating.
- Some foods have their own natural covering and no additional is needed. Potatoes and squash need to be pierced to prevent steam build-up and a possible explosion. Corn on the cob can be microwaved in its husk. Elevate these foods on a rack for better air circulation and more even cooking.

FROZEN CONVENIENCE FOODS

The microwave oven and frozen convenience foods make a perfect partnership for stopping hunger dead in its tracks. A dinner's journey from freezer to tummy can take as little as five minutes on "The Microwave Express."

Because virtually all frozen convenience foods have been precooked, it is not necessary to defrost them. They will defrost and reheat in one step. The cooking times on the packages we tested were quite reliable.

If no power level is given in the directions, assume that the manufacturer intends for you to use High power. Most of the packages we surveyed listed no power level.

One technique that we recommend was not mentioned on any packages we found. **You should stir, rearrange or turn over foods midway through cooking.** This promotes better results by more-evenly cooking the frozen areas of the food after the outer edges have defrosted and begun cooking.

Despite the advances made on package microwave cooking directions, you are still going to have to USE YOUR OWN INTELLIGENCE AND COOKING KNOWLEDGE to obtain best food results. Don't be reluctant to **open the oven door and touch, stir or rearrange the food.** If the package has said to microwave the food 8 minutes, and you can tell it is ready after 6 or 7 minutes, do not continue microwaving or you will overcook it!

Likewise, if you have microwaved food the prescribed 8 minutes, and it is not hot, add additional time. Don't berate the manufacturer. There are many variables involved, including: how solidly frozen the food is; whether it sat on your counter a few minutes before you had a chance to put it in the oven; and variances in electrical power within the city where you live.

WAX-PAPER WRAPPED BOXES

Method I: Place box on a paper towel or plate. (No need to unwrap box or pierce). **MW on High 5 to 6 minutes.** If you don't want to wash a serving dish, spoon onto plates right out of the box!

Method II: For frozen vegetables containing excess liquid (such as spinach or chopped broccoli), stand box in a 4-cup glass measure. Liquid will drain out the end of the box as food cooks. **MW on High 5 to 6 minutes.** Using a wooden spoon, press remaining liquid from box, and serve.

LARGE BAGS

To microwave a 20-ounce or larger bag of frozen vegetables, make a 1-inch slit on one side of bag. Place slit-side on paper towels or plate. Do not place bag directly on floor of microwave oven because dyes printed on the wrapper can stain it. **MW on High** according to chart on page 14. Turn bag over and redistribute contents midway through cooking.

BOILABLE POUCHES OR BAGS

Using a paring knife, make a large "X" almost from corner to corner through one side of the pouch. Place cut-side down in a microwave-safe serving dish. **MW on High 5 to 6 minutes.** To serve, lift hot pouch with tongs, and contents will empty through the "X."

FROZEN T.V. DINNERS

Many microwave oven manufacturers allow the use of aluminum foil "TV dinner trays" for microwaving the frozen food. There are two reasons we don't recommend doing this. First, since metal reflects microwaves, they cannot penetrate the bottom of the tray. The frozen foods will defrost and reheat only from the top; therefore, cooking is uneven.

Second, frozen dinners will look more appealing and cook evenly if microwaved on a dinner plate. It is extremely easy to transfer the frozen food from its foil tray to a plate: turn the tray upside down and pop the food out as if it were ice cubes. Cover plate with a "Micro Cover," see page 16. **Stirring or rearranging food midway through cooking, MW on 70% (Medium-high) until heated through.**

Some frozen dinners are packaged on a **microwave-safe plate** covered with foil, topped by a vented plastic lid. To heat in the microwave oven, remove the foil, replace the plastic lid and **MW on High,** according to package directions. Frozen foods packaged in paperboard trays and dishes are microwave safe. Microwave as package directs.

Entrees packaged in **deep rectangular foil dishes** should be removed from them for microwaving. Microwave-safe "frozen food dishes," such as those manufactured by Anchor Hocking and Rubbermaid, are just the right size for heating these entrees.

FROZEN BREADED FISH

Batter-coated foods are not as good when microwaved as when baked in a conventional oven. Fish sticks and fillets are quite acceptable when microwaved, although they aren't as crunchy. Do not defrost breaded fish products before microwaving. Breaded foods which required deep-fat frying will not cook successfully in the microwave oven.

Place the breaded fish item directly from the freezer onto a microwave meat or bacon rack. This allows air to circulate beneath during microwaving. So that coating will stay crisp, do not cover. Fish sticks and fillets come in quite a variety of package sizes. For a 12-ounce package (16 to 18 sticks), **MW on High 5 to 5½ minutes,** rearranging fish and rotating rack midway through cooking. For 8 (2-ounces each) fish fillets, **MW on High 5½ to 6½ minutes,** rearranging fish and rotating rack midway through cooking.

FROZEN FRIED CHICKEN

Precooked fried chicken should be microwaved on a meat or bacon rack. Arrange with meatiest portions to the outside of the rack. Cover with paper towel and **MW on High about 1½ to 2 minutes per piece,** or until heated through. Rearrange pieces midway through cooking. Do not microwave more than ten pieces at a time for best results.

FROZEN PIZZA

Any brand of frozen pizza microwaves best on a preheated microwave browning tray or "Pizza Crisper." **MW on High 6 to 7 minutes** for a 10-inch frozen pizza. You can also microwave pizza on a meat or bacon rack, although it won't be as crisp.

FROZEN CONVENIENCE FOOD CHART

ITEM	SIZE	POWER	MINUTES	SPECIAL INSTRUCTIONS
Appetizers	12 bite-size	100%	2½ to 4½	Place on paper towels on a meat or bacon rack.
Dinner Entrees	8 to 9-oz.	70%	6 to 8	Leave food in MW-safe containers. If food is in aluminum foil containers, remove to a MW-safe dish.
Pouch Dinners	5 to 6-oz.	100%	4 to 5	Cut an X through one side of pouch from corner to corner. Place cut side down in dish.
TV Dinners	7 to 9-oz. 11-oz. 17-oz.	70%	6 to 7 8 to 10 11 to 13	Leave food in MW-safe containers. If food is in aluminum foil tray pop out onto dinner plate. Cover. Rearrange midway through cooking.
Fried Chicken	per piece 16-oz. 32-oz.	100%	1½ to 2 6 to 7 12 to 14	Line meat or bacon rack with paper towels. Arrange single layer of chicken with meatier pieces toward the edge. Turn chicken over and rotate rack midway through cooking.
Pizza	7½-oz. 10.8-oz.	100%	4½ to 5 6 to 7	Place on preheated browning tray for crisper crust; or place on meat or bacon rack.
Fish Fillets	1 (2 to 3-oz.) 6-oz.	100%	1 to 1½ 3½ to 4½	Place on meat or bacon rack. Do not cover. Rearrange fish and rotate rack midway through cooking.
Fish Sticks	8-oz. 12-oz.	100%	3½ to 4½ 5 to 5½	
Vegetables (wax paper-wrapped box)	8 to 10 oz.	100%	5 to 6	Place unwrapped box on paper plate or towel; or stand box on end in measuring cup.
Vegetables in Pouch	8 to 10-oz.	100%	5 to 6	Cut an X through one side of pouch from corner to corner. Place cut side down in dish.
Vegetables in Large Bags	16-oz. 20-oz. 24-oz. 32-oz.	100%	8 to 10 10 to 12 12 to 14 14 to 16	Slit bag; place cut side down in dish. Redistribute contents midway through cooking.
Corn-on-the-Cob (whole ears)	1 2 4	100%	3½ to 4½ 6 to 7 11 to 12	Place in rectangular dish and cover with plastic wrap.
Corn-on-the-Cob (Short ears)	1 2 4	100%	2 to 2½ 4 to 5 6 to 7	Place in rectangular dish and cover with plastic wrap.
Breakfast Entrees	7-oz.	100%	2	Leave food in MW-safe containers. Rotate midway through cooking.
Pancakes	3 (1-oz.)	100%	1¼ to 1¾	Stack pancakes on a MW-safe plate. Do not cover.
Donuts Danish Honey Buns Muffins	4 (2-oz.) 1 (2-oz.) 1 (2-oz.) 1 (1¼-oz.) 4 (9-oz.) 1 (2¼-oz.) 6 (11.5-oz.) 1 (2-oz.)	100% 100% 100% 100% 100% 100% 100% 100%	1½ 30 to 35 sec. 15 to 17 sec. 20 sec. 1½ 35 to 40 sec. 3 30 to 35 sec.	Place on MW-safe plate. Do not cover.

CASSEROLE TECHNIQUE

Most casseroles consist of ingredients which have already been cooked. The final step is essentially REHEATING. Casseroles can be assembled ahead of time and refrigerated or frozen for later use.

Casseroles in the freezer are like money in the bank. Zap them in the microwave and make out like a bandit!

Frozen casseroles are so handy for easy summertime living. If you have a weekend vacation home, you can take them with you in the car. Defrost and reheat at your destination, and a casserole becomes an effortless meal.

During busy times or when taking a meal to a friend, casseroles are great. When we are out of town teaching microwave cooking classes, we frequently leave them for our families.

If you don't want the bother of retrieving a glass casserole dish, look for those made of disposable paperboard. These are safe for the microwave oven and can be thrown away after use.

Here are some helpful hints for casserole preparation.

- Unless you want the onion, celery or bell pepper in a casserole to remain crunchy, microwave them before adding to other ingredients. Place in a glass measure, cover with plastic wrap, and microwave on High 2 minutes per cup, or to desired softness.

- To convert conventional casserole recipes, liquid may need to be reduced.

- Casseroles will heat more evenly if they are placed in a round utensil and are covered.

- Stirring the contents midway through cooking also speeds the heating of a casserole. However, some casseroles, such as lasagne, can't be stirred. Shield the corners so they do not overcook. Cut four (3x1-inch) pieces of aluminum foil and tape around the outside corners of rectangular or square casseroles, using masking or transparent tape. The aluminum foil will reflect microwaves away from the corners.

- Crumb toppings should be added after stirring or rotating casserole midway through cooking. Do not cover unless you want soggy crumbs!

- Cheese toppings should be added at the very end of cooking time. If cheese has been put on top of casserole at the beginning of microwaving, it will become tough and rubbery. Use 70-percent power for casseroles containing cheese.

- When freezing casseroles, put crumb or cheese toppings in a separate plastic bag and tape it to packaging. They should be added only after frozen casserole has been defrosted and reheated.

- See page 101 for Casserole Freezing Directions.

REHEATING FOODS

Reheating leftovers in the microwave oven—everybody does it! Super easy and great results, right? Not always true! If you've had your piece of quiche become a melted mess of cheese, peas pop off the plate, or dinner rolls double for golf balls, then perhaps you can benefit from some reheating tips.

There are several factors to take into consideration when reheating foods. **Foods that are dense (potatoes) will take longer to heat than porous items (bread products.)** Microwave energy is attracted to foods that have a concentration of fat or sugar molecules. For instance, a sweet roll gets hotter more quickly than a plain dinner roll.

By giving some attention to the container and arrangement of the food, you can achieve better and more uniform reheating. Use a microwave-safe utensil for reheating. Soft plastics, such as Tupperware or whipped topping containers, will distort from the heat of the food. Fat and sugar in foods attract microwaves and become so hot they can "pit" soft plastic utensils.

Your first inclination when reheating foods is to pop the food item in the microwave oven and zap it with High power for fast results. High power can be so intense that foods will overcook and become tough (meats) or break down (quiche).

BEST POWER LEVELS FOR REHEATING

HIGH POWER (100%): Liquids; vegetables; water-based soups; pasta and rice; and bread products.

MEDIUM-HIGH POWER (70%): Meats; quiche; cream-based soups; casseroles; plated leftovers; and pies.

MEDIUM (50%): Layered casseroles which cannot be stirred.

PLATED LEFT-OVERS

- Arrange dense foods to the outside of the plate and the porous foods to the inside. For instance, cut a piece of swiss steak in half and place at opposite edges of the plate. In the center, place beans and mounded mashed potatoes, which should be indented. Butter or gravy can be placed in the "well." By distributing the food in this manner, all food will be heated evenly.

- We recommend using a plastic dome-shaped utensil such as a "Micro Cover" (manufactured by the American Family Scale Company) to cover the plated food for reheating. The cover won't touch the food since it rests on the rim of the plate or floor of the microwave oven.

- Reheat left-overs on 70% (Medium-high) 2 minutes per plate.

- Wrap bread products in a paper towel or cloth napkin to hold in heat. Before reheating, butter rolls or bread products that are to be eaten buttered. The melted butter will help hold the heat in the rolls longer.

APPETIZERS and BEVERAGES

NUTS

Nuts are one of man's oldest foods. They are a good source of protein and minerals. In recipes, different nuts may be subsituted for one another. Although peanuts are generally thought of as nuts, they are related to beans and peas, but can substitute for nuts. Chestnuts have been used since ancient times and are the only nuts which are actually eaten as vegetables.

Learn to prepare nuts in your microwave oven. They can be shelled, blanched, roasted or toasted.

When buying nuts unshelled, look for nuts which are free of cracks with a clean shell. One pound of nuts in the shell will yield 1 to 2 cups of nutmeats. **To make shelling easier,** place 2 cups of nuts in 1 cup of water. Cover and **MW on High 3 to 4 minutes.** Don't let them cool too long before cracking, or they will harden again.

To blanch nuts, place 1 cup water in a 4-cup glass measure. **MW on High 2½ to 3 minutes,** or until boiling. Add ½ to ¾ cup nuts. **MW on High 30 to 60 seconds.** Drain and rub nuts between two paper towels to remove skin.

To roast nuts, place 1 cup nuts in a single layer in a 9-inch glass pie plate. For peanuts, pecans and walnuts, **MW on High 3 to 4 minutes,** stirring every minute. For almonds, cashews, filberts and pine nuts, **MW on High 9 to 11 minutes,** stirring every minute after 4 minutes. If desired, add 1 to 2 tablespoons margarine and seasoning. Stir until margarine is melted and nuts are coated evenly. Cool and store in an air-tight container.

To roast chestnuts, use a sharp knife to make a cut on the flat side of each nut. Place approximately 20 chestnuts in a single layer in a 9-inch glass pie plate. Stirring every minute, **MW on High 3 to 4 minutes,** or until nuts are softened. Remove shells, peel skin and eat warm.

SNOWY SPICED NUTS

- 4 tablespoons butter or margarine
- 2 cups mixed nuts
- 1¼ cups powdered sugar
- 1 teaspoon cinnamon
- 1 teaspoon nutmeg
- ½ teaspoon allspice or cloves

Place butter in a 9-inch glass pie plate and **MW on High 30 seconds,** or until melted. Add nuts and stir to coat evenly. Stirring midway through cooking, **MW on High 4 to 5 minutes,** or until butter is absorbed and nuts are roasted.

Combine sugar, cinnamon, nutmeg and allspice in a paper bag. Add nuts and shake to coat nuts. Let cool on a plate. Serve in a candy dish or bowl. Yields 2 cups.

POPCORN

We absolutely love making popcorn in the microwave oven! In merely three or four minutes, you can turn out a batch with no oily mess or residual heat as in conventional methods.

Popcorn is a true all-American snack loved by kids and adults alike. Indians first discovered the variety of edible corn which will "pop." The kernels come from small ears of corn about six-inches long, which go through a drying process. Heat causes the moisture and air in the dried kernel to explode inside-out to a size many times larger.

Since popcorn has so little moisture, kernels have to become extremely hot in order to pop. **DO NOT MICROWAVE POPCORN IN A PAPER BAG!** Many paper bags are made from recycled materials, including bits of metal, that can catch on fire with the extreme high heat of the kernels. However, prepackaged microwave popcorn can be found in your supermarket. It is packaged in its own specially-designed white paper bag for popping. Follow package directions.

For microwaving popcorn we recommend a cone-shaped popper especially made for the microwave oven. Don't use casseroles or other microwave-safe utensils because they can break from the heat of the kernels. Follow the manufacturer's recommendations for your particular microwave oven. Popping corn improperly is a major cause of service calls on microwave ovens.

Oil is not used in most microwave poppers. Do not add salt before popping, as it distorts the microwaves.

For popcorn kernels to pop well, they must contain the proper moisture content. If your kernels are too dry, place them into an air-tight container along with a damp paper towel and seal for a few days. You can also add one tablespoon of water per quart of kernels and shake daily for a few days. Storing unpopped kernels in the refrigerator or freezer helps maintain the proper moisture content.

Appetizers

PARMESAN POPCORN

- 2 quarts popped popcorn
- 4 tablespoons margarine
- 1 teaspoon garlic salt
- 1 teaspoon Italian seasoning
- 2 tablespoons grated parmesan cheese

Place popcorn into a serving bowl. Place margarine in a 1-cup glass measure and **MW on High 40 seconds,** or until melted. Add garlic salt and herbs; stir well. Drizzle over popcorn and toss well. Sprinkle with cheese and toss again. *Best served warm, but can be stored in an airtight container.* Yields 2 quarts.

CARAMEL CORN

This recipe works best with freshly-popped warm popcorn.

- 2 quarts popped popcorn
- ½ cup packed brown sugar
- ¼ cup margarine
- 2 tablespoons dark corn syrup
- ¼ teaspoon salt
- ¼ teaspoon baking soda
- ½ teaspoon vanilla

Place popcorn into a 4 to 6-quart microwave-safe utensil. Combine sugar, margarine, syrup and salt in a 1-quart glass batter bowl. **MW on High 1½ minutes;** stir. **MW on High 2 to 2½ minutes.** Add baking soda and vanilla; stir well.

Pour mixture immediately over popcorn, stirring to coat evenly. **MW on High 2 minutes,** stirring midway through cooking. Turn out onto a piece of waxed paper or aluminum foil to cool. Break into pieces and store in an airtight container. Yields 2 quarts.

---

*"Yours is such a great cookbook, I have to keep **giving** them to people, as I can't part with my copy long enough to just **loan** it to them! I enjoy mine daily, and have made about 75% of the recipes at least once—usually each new one I try turns into a favorite."*

Loren Rice, Brighton, Massachusetts

➡ *Appetizers*

NUTTY POPCORN BALLS

1 (14-ounce) bag caramels
3 tablespoons milk
2 quarts popped popcorn
1 cup dry-roasted salted peanuts

Combine caramels and milk in a 1-quart glass batter bowl. **MW on 70% (Medium-high) 5 to 6 minutes,** or just until bubbly. Stirring once midway through cooking, pour mixture over popcorn and peanuts in a mixing bowl, tossing to coat evenly. Butter hands and form mixture into balls. Place on waxed paper or foil to cool. Yields 18 (3-inch) balls.

TEXAS TRASH

2 quarts popped popcorn
1 (10-ounce) box extra-thin pretzel sticks
2 cups Chex cereal squares
1½ cups dry-roasted peanuts or other nuts (preferably unsalted)
½ cup margarine
1 teaspoon garlic salt
1 teaspoon onion salt
1 teaspoon celery salt
2 tablespoons worcestershire sauce

Place popcorn, pretzels, cereal and nuts in a 5 to 6-quart microwave safe utensil. Set aside.

Place margarine in a 2-cup glass measure and **MW on High 1 minute,** or until melted. Add seasonings and worcestershire sauce; stir well. Drizzle over popcorn mixture and toss to coat well.

MW on High 4 minutes, stirring once midway through cooking. Let cool and store in an airtight container. Yields 3 quarts.

Quick-Tip ➡ When purchasing packaged nuts already shelled, 8 ounces equals about 1½ to 2 cups of nutmeats.

Appetizers

MUSHROOMS VERMOUTH

¼ cup olive oil
1 clove garlic, minced
¼ cup vermouth
1 teaspoon Italian seasoning
1 pound fresh mushrooms, cleaned

Combine olive oil and garlic in a 1-cup glass measure. Cover and **MW on High 2 minutes.** Add vermouth and seasoning. Place mushrooms in a 1-quart casserole. Pour margarine mixture over mushrooms. Cover and **MW on High 3 minutes.** Do not overcook mushrooms or they will shrink due to their high moisture content. Serve with toothpicks. Yields about 40 mushrooms.

NOTE: *May be refrigerated and served chilled.*

MANITARIA PARAYEMISTA

Greek stuffed mushrooms

1 (10-ounce) package frozen spinach
½ cup grated parmesan cheese
4 ounces feta cheese, rinsed and crumbled
½ cup finely-chopped green onions
½ cup finely-chopped fresh parsley
1 pound fresh mushrooms, cleaned and stemmed (about 24)

Stand box of spinach in a 4-cup glass measure. **MW on High 3 to 4 minutes,** or just until defrosted. Press liquid from box and discard. Transfer spinach to same measuring cup. Add cheeses, onions and parsley. Fill mushroom caps with spinach mixture, mounding high in center.

Arrange mushrooms on a 12-inch round glass or plastic tray, leaving center of the tray empty. Rotating tray midway through cooking, **MW on High 8 to 10 minutes,** or until mushrooms are desired doneness. Yields approximately 24 stuffed mushrooms.

Appetizers

CURRY DIP

2 tablespoons finely-chopped onion
1 to 2 teaspoons curry powder
1 cup sour cream

Place onion in a 1-cup glass measure. Cover and **MW on High 1½ to 2 minutes.** Blend onion, curry powder and sour cream together. Refrigerate. *Serve with assorted fresh vegetables.* Yields 1 cup.

ALMOND CHICKEN DIP

1 (8-ounce) package cream cheese
1 teaspoon worcestershire sauce
⅛ teaspoon garlic powder
1 (4-ounce) can mushroom pieces, including liquid
1 (5-ounce) can boned chicken
½ cup sliced almonds

Place unwrapped cream cheese in a 1-quart casserole. **MW on 50% (Medium) 1½ to 2 minutes,** or until softened. Blend in worcestershire sauce and garlic powder. Add mushrooms with liquid, chicken and almonds. **MW 3 minutes on 50%,** or until heated through. *Serve warm with crackers or melba rounds.* Yields 3 cups.

SWEDISH SAUSAGE DIP

1 pound bulk sausage
1 (4-ounce) can mushroom pieces, drained
1 cup plain yogurt
1 cup sour cream
3 green onions with tops, thinly sliced

Crumble sausage into a hard-plastic colander set in a 1½-quart round casserole. **MW on High 5 to 6 minutes,** stirring midway through cooking. Discard grease and transfer sausage into same casserole. Add mushrooms, yogurt and sour cream. **MW on 70% (Medium-high) 3 minutes,** or until heated through. Top with onions. *Serve with assorted crackers.* Yields 1 quart.

Appetizers

CHILE CON QUESO

1 (2 pound) carton Velveeta cheese
1 (10-ounce) can Rotel tomatoes and green chilies

Cube cheese and combine with tomatoes and green chilies in a 1-quart casserole. Stirring twice during cooking, **MW on 70% (Medium-high) 6 minutes,** or until cheese melts. Yields 3 cups.

> **Variation:** "CHILI" CON QUESO. Substitute 1 (16-ounce) can chili without beans for the tomatoes and green chilies.

SAUSAGE CON QUESO

In honor of our WACO TRIBUNE-HERALD

1 pound bulk sausage
1 (2-pound) carton Velveeta cheese, cut into 1-inch cubes
1 (10-ounce) can Rotel tomatoes and green chilies

Crumble sausage into a hard-plastic colander set in a 2-quart round casserole. Cover with paper toweling. **MW on High 6 to 7 minutes,** stirring midway through and at end of cooking. Discard grease. Set sausage aside.

Combine cheese and tomatoes in same casserole; cover. Stirring midway through cooking, **MW on 70% (Medium-high) 7 to 8 minutes,** or until cheese melts. Stir in sausage. **MW on 70% 2 to 2½ minutes,** or until heated through. *Serve with tortilla chips.* Yields 1½ quarts.

AVOCADO SAUSAGE SQUARES

1 pound bulk sausage
1 cup shredded Monterey Jack cheese (4 ounces)
1 avocado, chopped (tossed with lemon juice)
2 green onions with tops, thinly sliced
4 eggs
½ cup milk
½ cup crushed flavored tortilla chips

Crumble sausage into a hard-plastic colander set in an 8x8x2-inch glass dish. **MW on High 6 to 7 minutes,** stirring once midway through and at end of cooking. Discard grease. Spread sausage evenly over bottom of dish. Sprinkle cheese on top, then avocado, then onions. Lightly beat together eggs and milk; pour over all. Sprinkle crushed chips on top. Rotating midway through cooking, **MW on 70% (Medium-high) 11 to 13 minutes,** or until set. Let stand 5 minutes. Makes 9 servings.

Appetizers

MEXICAN PIZZA

- 1 pound ground beef
- 1 (1¼ ounce) package taco seasoning
- ½ cup water
- 1 (16-ounce) can refried beans with chilies
- ½ cup taco sauce
- ½ cup shredded cheddar cheese
- 1 (6-ounce) package frozen avocado dip, thawed
- 1 tomato, chopped

Crumble beef into a hard-plastic colander set in a 2-quart round casserole. **MW on High 5 to 6 minutes,** stirring midway through cooking. Discard grease and transfer beef into same casserole. Stir in taco seasoning and water. Cover and **MW on High 5 minutes.**

Combine beans and taco sauce and spread evenly over a 12-inch round glass or plastic tray. Top with meat mixture. **MW on High 5 to 6 minutes,** rotating tray midway through cooking.

Sprinkle cheese over "pizza." Spoon avocado dip in center and distribute tomato over all. Makes 12 servings.

Serving Suggestion: *Serve as a dip with large-size corn chips. Homemade guacamole can be substituted for the avocado dip.*

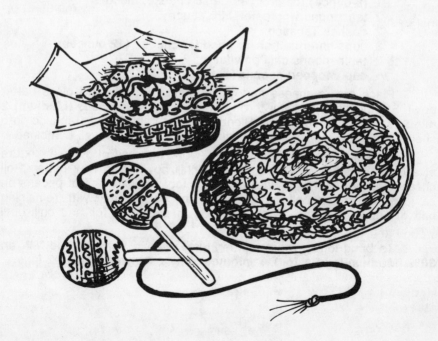

Appetizers

BISCUIT BITS

- 4 tablespoons margarine
- 1 (8-ounce) can refrigerated biscuits (10 biscuits)
- ½ cup shredded extra-sharp cheddar cheese
- 2 tablespoons imitation bacon bits
- ½ teaspoon chili powder
- Chopped green chilies (optional)

Place margarine in an 8-inch round glass cake pan. **MW on High 40 seconds,** or until melted. Cut each biscuit into four pieces. Stir biscuits into melted margarine, making sure that each piece is well coated. Arrange in a single layer in same dish. **MW on High 2 minutes,** rotating dish after 1 minute.

Sprinkle with cheese, then bacon bits. Sprinkle with chili powder (and green chilies, if desired). **MW on High 1 to 1½ minutes.** Let stand 2 minutes before serving. Yields 40 biscuit bits.

CHILI PECAN LOG

- 1 (8-ounce) package cream cheese
- 1 (6-ounce) roll process garlic cheese, diced
- 2 teaspoons worcestershire sauce
- 2 dashes Tabasco
- 3 cups shredded sharp cheddar cheese (12 ounces)
- 1 to 2 tablespoons chili powder
- ½ cup chopped pecans

Place cream cheese and garlic cheese in a 2-quart glass batter bowl. Stirring once midway through cooking, **MW on 50% (Medium) 2 minutes,** or until softened. Blend worcestershire and Tabasco into cheeses. Add shredded cheddar cheese and stir until well combined.

To make into two (6-inch) logs, work with half of cheese mixture at a time. Sprinkle desired amount of chili powder on wax paper. Roll cheese mixture in chili powder to form a log. Sprinkle half of pecans on same waxed paper; roll log in pecans to coat. Repeat with remaining half of cheese mixture. Wrap each log in aluminum foil and chill until firm.

To bring log to room temperature, remove from foil and **MW on 30% (Medium-low) 1 to 1½ minutes.** *May be frozen.* Yields 2 logs.

Appetizers

KAOLA PINEAPPLE-HAM KABOBS

- 1 (8-ounce) can pineapple chunks, drained
- 1¼ pounds fully cooked ham, cut into 1-inch cubes
- 2 medium zucchini, cut into 1-inch pieces
- 4 ounces fresh mushrooms
- ¾ cup apricot-pineapple preserves
- 1 teaspoon Dijon-style mustard
- 1 teaspoon soy sauce
- 16 cherry tomatoes

Alternate pineapple, ham, zucchini and mushrooms on bamboo skewers, leaving room at each end for a cherry tomato. Arrange on a meat rack or in a 2-quart rectangular dish. Cover with plastic wrap and **MW on High 4 minutes,** rotating dish midway through cooking. Turn kabobs over, top to bottom, and pour off liquid that has collected.

Combine preserves, mustard and soy sauce. Brush glaze over top of kabobs. Cover and **MW on High 4 minutes,** rotating dish midway through cooking. Turn kabobs over, top to bottom; add cherry tomatoes at each end of skewer. Brush glaze over top of kabobs. Cover and **MW on High 1 to 1½ minutes,** or until heated through. Yields 8 kabobs.

Appetizers

HOT CRABMEAT ELEGANTE

1 (6-ounce) roll process garlic cheese, diced
1 (10 ¾-ounce) can cream of mushroom soup
1 (6-ounce) can crabmeat, drained
2 tablespoons sherry

Combine cheese with soup in a 1½-quart casserole. **MW on 70% (Medium-high) 3 to 4 minutes,** stirring midway through cooking. Add crabmeat and sherry. **MW on 70% 2 minutes,** or until heated through. Yields 1 quart.

Serving Suggestion: *Keep warm in a chafing dish and serve with assorted crackers.*

SHRIMP BUTTER

1 (8-ounce) package cream cheese
4 tablespoons margarine
¼ cup minced onion
4 tablespoons mayonnaise
1 tablespoon lemon juice
1 tablespoon snipped fresh parsley
½ teaspoon garlic salt
2 cups finely chopped cooked shrimp

Combine cream cheese and margarine in a 4-cup glass measure. **MW on 50% (Medium) 1½ to 2 minutes,** or until softened but not melted. Stir in onion, mayonnaise, lemon juice, parsley and garlic salt; blend well. Add shrimp; stir. Refrigerate overnight. *Serve with fresh vegetables or crackers.* Yields 2½ to 3 cups.

Quick-Tip ➡ Use your microwave oven to soften food quickly which should be of a spreadable consistency (such as frostings, cheese spreads, garlic butter, etc.). **MW on 30% (Medium-low).**

Beverages

CHRISTMAS SPICE

- 1 piece stick cinnamon
- 1 teaspoon cinnamon
- 1 teaspoon whole allspice
- ½ teaspoon whole cloves
- ½ teaspoon ginger
- ¼ teaspoon nutmeg

To make one packet, combine above ingredients in the center of a 6-inch square of cheesecloth. Gather corners together to form a bundle. Tie bundle with string or dental floss.

Place packet in a 2-cup glass measure filled with hot tap water. **MW on High 5 to 6 minutes,** or until water is boiling. Continue to **MW on 30% (Medium-low)** to send fragrance of Christmas Spice throughout house. Additional water may need to be added if microwaved for an extended time.

CRANBERRY WASSAIL

- 1 quart cranberry juice cocktail
- 1 quart pineapple juice
- ½ cup sugar
- 1 Christmas Spice packet

Combine all ingredients in a 2-quart batter bowl. Cover with plastic wrap. Stirring twice during cooking, **MW on High 14 to 15 minutes,** or until boiling. Continue to **MW on 50% (Medium) 4 to 5 minutes** for spices to flavor wassail. Yields 2 quarts.

Beverages ←

INDIVIDUAL HOT CHOCOLATE

1 mug of milk
1 tablespoon chocolate syrup or 2 teaspoons instant cocoa mix
Marshmallows

Combine milk and chocolate in your favorite microwave-safe mug. **Set temperature probe for 150° and MW on 70% (Medium-high).** Top with marshmallows, if desired.

To make multiple servings, combine ingredients in a large glass measure and program as above.

Variation: COCOA CADILLAC – Add 2 tablespoons Galliano liqueur.

KAHLUA COCOA – Add 2 tablespoons Kahlua liqueur.

COCOA GRASSHOPPER – Add 1 tablespoon Creme de Menthe and use peppermint stick as stirrer.

HOMEMADE COCOA MIX

16 ounces instant dry milk (4 cups)
1½ cups sugar
¾ cup cocoa
½ teaspoon salt

Mix dry milk, sugar, cocoa and salt together in a glass jar. To store, keep jar tightly covered. Reconstitute as follows; **MW on High** to desired temperature.

AMOUNT OF MIX	AMOUNT OF WATER	YIELD
3 level tablespoons	5 ounces water	1 (5-ounce) coffee cup
4 level tablespoons	8 ounces water	1 (8-ounce) mug
1 cup	4 cups water	1 quart
1 recipe of mix	4 quarts water	1 gallon

> Beverages

SLEEPYTIME SWISS MOCHA MIX

No caffeine to keep you awake!

- ¾ cup sugar
- ⅓ cup cocoa
- ½ cup instant decaffeinated coffee powder
- 1 cup instant non-dairy creamer

Blend sugar and cocoa until no lumps remain. Stir in coffee powder and creamer. Store in an airtight jar. Reconstitute as follows; **MW on High** to desired temperature.

AMOUNT OF MIX	AMOUNT OF WATER	YIELD
4 level teaspoons	5 ounces water	1 (5-ounce) coffee cup
2 level tablespoons	8 ounces water	1 (8-ounce) mug
½ cup	4 cups water	1 quart
1 recipe of mix	4 quarts water	1 gallon

Variation: MEXICAN MOCHA Add 1 tablespoon coffee liqueur (such as Kahlua) to each (8-ounce) mug, and top with whipped cream. Dust with cinnamon.

Beverages

COZY ROSÉ TEA

- 2 cups water
- 4 regular-size tea bags
- ½ cup sugar
- 1 fifth rose wine (26 ounces)
- 1 (6-ounce) can frozen lemonade concentrate

Measure water into a 2-quart glass batter bowl. **MW on High 4 minutes,** or until boiling. Add tea bags and steep for 5 minutes. Remove tea bags and stir in sugar until dissolved. Add wine and lemonade concentrate. **Set temperature probe for 160° and MW on High.** Yields about 1½ quarts (8 servings - 6 ounces each).

NOTE: *Can be made in advance and refrigerated prior to reheating.*

LOW-CALORIE SPICED TEA MIX

30 calories per 8-ounce serving

- ¾ cup lemon iced tea mix sweetened with Nutra-Sweet™ or artificial sweetener
- ½ cup instant orange drink mix (such as Tang)
- ½ teaspoon cinnamon
- ¼ teaspoon cloves

Combine iced tea mix, orange drink mix, cinnamon and cloves. Store in an airtight jar. Reconstitute as follows; **MW on High** to desired temperature.

AMOUNT OF MIX	AMOUNT OF WATER	YIELD
2 level teaspoons	5 ounces water	1 (5-ounce) coffee cup
1 level tablespoon	8 ounces water	1 (8-ounce) mug
Scant ⅓ cup	4 cups water	1 quart
1 recipe of mix	4 quarts water	1 gallon

SOUPS and SALADS

Soups

GAZPACHO de ANN

½ cup EACH chopped celery, onion and green bell pepper
1 (46-ounce) can tomato juice
1 cup chopped tomatoes
3 tablespoons wine vinegar
2 tablespoons olive oil
1 garlic clove, minced
2 teaspoons snipped parsley
1 teaspoon worcestershire sauce
1 teaspoon salt
¼ teaspoon fresh ground pepper

If you prefer your celery, onion and green pepper less than crisp, put them in a 2-cup glass measure. Cover with plastic wrap and **MW on High 3 minutes.** Combine with remaining ingredients and pour into glass jars. Cover and refrigerate overnight so that flavors can blend. Serve cold. *Will keep for 2 weeks.* Yields 2 quarts.

COLD SPINACH SOUP

Also good served hot!

1 (10-ounce) package frozen chopped spinach
3 tablespoons butter or margarine
½ cup finely-chopped onion
2½ tablespoons flour
2 teaspoons instant chicken bouillon granules
⅛ teaspoon nutmeg
3½ cups milk
Salt and pepper to taste
Sour cream and nutmeg for garnish (optional)

Place unopened package of spinach in a 2-quart batter bowl. **MW on High 4 to 5 minutes.** Squeeze remaining water out of spinach and discard liquid. Set spinach aside.

Combine butter and onion in same 2-quart batter bowl. Cover with plastic wrap and **MW on High 4 minutes.** Stir in flour, bouillon and nutmeg. Using a whisk, blend in milk. Whisking every 2 minutes, **MW on High 8 minutes,** or until thickened. Add spinach, salt and pepper to taste. Chill. If desired, serve with a dollop of sour cream dusted with nutmeg. Yields 1 generous quart.

SWISS CHEESE VEGETABLE SOUP

1 (1 ⅝-ounce) package Knorr Swiss Vegetable Soupmix
3 cups hottest tap water
1 cup milk or light cream
1½ cups shredded Swiss cheese (6 ounces)

Combine soupmix and water in a 2-quart glass batter bowl. Cover with plastic wrap and **MW on High 15 minutes.** Stir in milk and cheese. Re-cover and **MW on 70% (Medium-high) 5 minutes,** or until heated through. Yields 1 generous quart.

FRENCH ONION SOUP

1½ pounds onions (about 4 medium)
4 tablespoons butter
4 (10½-ounce) cans condensed beef broth
1 cup water
½ cup dry white wine or Vermouth (optional)
½ teaspoon celery salt
¼ teaspoon garlic powder
French bread slices, toasted
Grated parmesan cheese

Peel onions and halve lengthwise. Slice thinly in food processor or by hand. Place onions in a 4-quart simmer pot along with butter; cover. Stirring midway through cooking, **MW on High 10 minutes,** or until onions are transparent.

Add broth, water, wine, celery salt and garlic powder. Cover and **MW on High 10 minutes,** or until onions are tender and soup is hot. Top each serving with French bread sprinkled with cheese. Yields 2 quarts.

Quick-Tip ➡ Plan in advance to make soup the day before you need it—flavors will have more time to blend, and dinner will be ready in the time it takes to reheat your soup.

Soups

BAKED POTATO SOUP

- 1 (2-pound) bag frozen Southern-style hash-brown potatoes
- ⅔ cup finely chopped onion
- 4 tablespoons margarine
- 1 (14½-ounce) can condensed chicken broth
- 1½ cups milk
- ¼ teaspoon white pepper
- Sour cream
- Shredded cheddar cheese
- Chives or thinly-sliced green onion tops
- Crumbled bacon (optional)

Place bag of frozen potatoes on paper plate. **MW on High 15 to 17 minutes,** turning bag upside down and redistributing contents every 5 minutes.

Combine onion and margarine in a 3-quart casserole. Cover and **MW on High 3 to 3½ minutes.** Place potatoes, onion and broth in food processor or mixer. Process until desired consistency.

Pour potato mixture into 3-quart casserole. Blend in milk and pepper; cover. Stirring midway through cooking, **MW on High 4 to 5 minutes,** or until heated through. Serve in bowls or mugs. Garnish top of individual servings with sour cream, cheese, chives and bacon. Yields 2 quarts.

CREAM OF PUMPKIN SOUP

- 2 tablespoons butter or margarine
- 1 large onion, finely minced or pureed
- 1½ teaspoons flour
- 1 (10½-ounce) can condensed chicken broth
- 1 (16-ounce) can pumpkin
- 2 cups milk
- 1 cup light cream
- 1 teaspoon salt
- ¼ teaspoon ginger
- ⅛ teaspoon pepper
- ⅛ teaspoon cinnamon

In a 3-quart casserole combine butter and onion. **MW on High 3 to 4 minutes.** Add flour; stir until blended. Using a whisk, add chicken broth. Whisking midway through cooking, **MW on High 7 to 8 minutes,** or until bubbly. Blend in remaining ingredients. Stir until smooth. **MW on 70% (Medium-high) 10 to 12 minutes,** or until hot. Serve immediately. Yields 2 quarts.

WILLIAMSBURG PEANUT SOUP

In honor of our RICHMOND NEWS-LEADER

- 4 tablespoons margarine
- ½ cup finely-chopped onion
- ¼ cup finely-chopped celery
- 3 tablespoons flour
- 2 (14½-ounce) cans chicken broth
- 1 cup smooth peanut butter
- 1 cup light cream or half and half
- ½ cup chopped roasted peanuts

Place margarine, onion and celery in a 2-quart batter bowl. Cover with plastic wrap and **MW on High 4 minutes.** Blend in flour. Using a whisk, blend in chicken broth. Whisking twice during cooking, **MW on High 10 to 12 minutes,** or until mixture comes to a boil.

Using whisk, blend in peanut butter and cream. **MW on 70% (Medium-high)** until heated to serving temperature. Do not boil, because cream will curdle. To serve, sprinkle bowl of soup with chopped peanuts. Yields 2 quarts.

AVGOLEMONO SOUPA

Greek chicken-rice soup with egg and lemon from our friend, Verna Gianopoulos

- 2 (10½-ounce) cans condensed chicken-rice soup
- 2 soup cans water
- 1 egg, well beaten
- 2 medium lemons

In a 2-quart bowl, combine soup and water. Cover and **MW on High 6 to 8 minutes.** *(If using temperature probe, set for 170°.)* Gradually add some of the hot soup mixture to the egg; stir until completely combined. Return mixture to the hot soup, stirring to combine.

MW 2 to 2½ minutes, or until soup thickens slightly. Squeeze the juice from one lemon, and thinly slice the other lemon. Just before serving, add lemon juice and slices to soup. **MW on High** to desired serving temperature. Yields 1½ quarts.

HEARTY CHICKEN-RICE SOUP

- 1 (2½ to 3-pound) fryer, cut up
- 1 onion, cut into chunks
- 1 rib celery, sliced
- 1 carrot, cut into sticks
- 1 clove garlic
- 1 bay leaf
- ½ teaspoon oregano
- 10 cups water
- 1 (10-ounce) package frozen peas and carrots
- 1⅓ cups instant rice
- Salt and pepper to taste

To stew chicken: Place chicken, onion, celery, carrot, garlic, bay leaf and oregano in a 4-quart simmer pot. Add water. Cover and **MW on High 30 to 35 minutes,** rearranging chicken pieces midway through cooking. Discard garlic and bay leaf.

Remove chicken pieces from pot and add frozen peas and carrots. Cover and **MW on High 10 minutes,** or until boiling. Add rice, salt and pepper; let stand until rice rehydrates (about 10 minutes). Meanwhile, debone chicken. Add to soup and reheat if necessary. Makes 10 to 12 servings.

Soups

BIG BIRD SOUP

- 1 quart turkey or poultry stock
- 2 cups diced leftover turkey
- 1 (16-ounce) can mixed vegetables including liquid
- 1½ cups leftover cooked rice
- ¼ teaspoon basil
- ¼ teaspoon pepper
- Pinch of allspice
- ¼ teaspoon sage
- ¼ teaspoon Kitchen Bouquet

Combine all ingredients in a 2-quart glass batter bowl. **MW on High 10 minutes,** or to desired temperature. *(If using temperature probe, set to 170°.)* Yields 2 quarts.

MEATBALL MINESTRONE

- 1 pound ground beef
- 1 egg
- ½ cup Italian-seasoned dry bread crumbs
- ¼ cup ketchup
- 1 medium onion, chopped
- 1 clove garlic, minced
- 2 tablespoons margarine
- 3 (10½-ounce) cans condensed beef broth
- 3 soup cans of water
- 1 (16-ounce) can sliced stewed tomatoes
- 1 bay leaf
- ¼ teaspoon basil
- ¼ teaspoon pepper
- 1 (10-ounce) package frozen mixed vegetables
- 6 ounces vermicelli, broken into 2" pieces (about 2 cups)

To make meatballs, combine ground beef, egg, bread crumbs and ketchup. Form into bite-size balls. Place on paper toweling in a 2-quart rectangular glass dish. **MW on High 6 to 7 minutes,** redistributing meatballs after 4 minutes. Set aside.

Place onion, garlic and margarine in a 4-quart simmer pot. Cover and **MW on High 3 minutes.** Add broth, water, tomatoes (including liquid), bay leaf, basil, pepper and frozen vegetables; stir. Cover and **MW on High 5 minutes.** Add vermicelli; stir. Cover and **MW on High 15 minutes,** stirring once midway through cooking. Add reserved meatballs and **MW on High 5 minutes**. Remove bay leaf before serving. Makes 8 servings.

DUCK GUMBO

1 (4-pound) duck, cut into pieces
Hot water
½ cup oil
½ cup flour
1 cup finely-chopped celery
1 cup finely-chopped onion
1 (16-ounce) can sliced stewed tomatoes
3 cups duck stock (reserved from stewing duck)
2 to 4 teaspoons Tony Chachere's Creole Seasoning
1 tablespoon filé powder
Cooked rice

TO STEW DUCK

Place duck pieces in a 4-quart simmer pot. Add hot tap water to cover duck. Cover and **MW on High 15 minutes; then MW on 50% (Medium) 30 minutes,** rearranging pieces midway through cooking.

While duck is still warm, remove meat and chop. Reserve stock and chill. Skim fat from top of stock before using.

TO MAKE BROWN ROUX

Blend oil and flour in a 4-cup glass measure. **MW on High 4 minutes;** stir. **MW on High 1 minute at a time,** stirring after every minute, until a deep brown roux is formed. Immediately add celery and onion; stir. **MW on High 6 to 7 minutes,** stirring midway through cooking.

Transfer roux to a 4-quart simmer pot. Put tomatoes, including liquid, into food processor and pulse until tomatoes are small pieces. Add to roux along with reserved stock, duck meat and Creole Seasoning. Cover and **MW on High 15 minutes.** Sprinkle with file powder; stir. Serve over rice. Makes 8 servings.

Quick-Tip ➡ When soups have been refrigerated, it is very easy to skim the fat which rises to the surface of the soup. If you have no time to wait, however, fat can be removed from the surface of hot soup by laying a paper towel on the surface of the soup. When it has absorbed as much fat as it will hold, remove it. Repeat if necessary.

SALADS

Salad from the microwave? We're not cooking the Romaine, but we do use the microwave where possible in salad making to keep kitchens cool in the summer.

Microwave owners sometimes don't think about using their ovens for small tasks which are part of conventional recipes. Some examples are: dissolving gelatin; sautéing vegetables; baking a pie shell to be filled with cold ingredients; or defrosting frozen ingredients.

To dissolve unflavored gelatin, first stir it into a cold liquid. This softens the gelatin. Next, microwave the mixture on high power until boiling to dissolve gelatin. Complete recipe as directed.

To prepare four-serving size flavored gelatins, place one cup of water in a 2-cup glass measure. **MW on High 2½ to 3 minutes,** or until boiling. Stir while adding gelatin and until dissolved. Immediately add ice cubes until the liquid level reaches the 2-cup mark. Stir until ice melts. Refrigerate and use as desired. By using ice to chill gelatin rapidly, it will be set in about 30 minutes. This quick-set method is also helpful when needed for a sick child.

WHITE AND BLUEBERRY SALAD

Also doubles as a light dessert

- 1 cup half & half
- ⅔ cup sugar
- 1 envelope unflavored gelatin
- ¼ cup cold water
- 2 cups sour cream
- 1 teaspoon vanilla
- 1 cup water
- 1 (3-ounce) package strawberry flavored gelatin
- 1 (13-ounce) can blueberries, including liquid

Combine half & half and sugar in a 4-cup glass measure. **MW on 70% (Medium-high) 1½ to 2 minutes,** or until sugar dissolves but DO NOT BOIL. Dissolve gelatin in cold water and add to half & half mixture. Add sour cream and vanilla; blend. Pour into an 8x8x2-inch dish. Refrigerate until set.

Place water in a 4-cup glass measure and **MW on High 2½ to 3 minutes,** or until boiling. Dissolve gelatin in water. Add can of blueberries, including liquid. Cool. Pour over white layer and refrigerate until set. Makes 9 servings.

NOTE: *One (10-ounce) package frozen strawberries may be substituted for blueberries.*

Salads

CUCUMBER SALAD

Delightfully delicate and cool

- ¾ cup water
- 1 (3-ounce) package lime flavored gelatin
- 1 medium cucumber
- 1 tablespoon minced fresh onion
- 1 cup mayonnaise
- 1 cup sour cream

Place water in a 4-cup glass measure and **MW on High 2 to 2½ minutes,** or until boiling. Stir in gelatin until dissolved. Cool to room temperature.

Slice cucumber in half and remove seeds. Grate cucumber and add to cooled gelatin along with onion, mayonnaise and sour cream. Pour into a 8x8x2-inch dish. Refrigerate until set. Makes 8 servings.

MOLDED GAZPACHO SALAD

A Spanish refreshment

- 2 envelopes unflavored gelatin
- 3 cups tomato juice
- ⅓ cup red wine vinegar
- Dash of Tabasco sauce
- 1 cup ground cooked ham
- 1 medium cucumber, peeled and diced
- ¼ cup chopped onion
- ¼ cup chopped celery

Sprinkle gelatin over 1 cup of tomato juice in a 4-cup glass measure. Blend well. **MW on High 2½ to 3 minutes,** or until boiling. Blend in remaining tomato juice, vinegar and Tabasco. Chill until partially set.

Lightly oil a 1½-quart mold. Add ham, cucumber, onion and celery to tomato juice mixture and pour into prepared mold. Refrigerate until set. Serve with Avocado Dressing, page 47. Makes 6 servings.

Salads

HAWAIIAN SLAW

- 2 envelopes unflavored gelatin
- 3 cups orange juice, divided
- 1 tablespoon lemon juice
- 1 (8¼-ounce) can crushed pineapple, undrained
- 1 carrot, grated
- 1 cup finely shredded cabbage

Sprinkle gelatin over 1 cup orange juice in a 2-quart batter bowl. Stirring midway through cooking, **MW on High 1½ to 2 minutes,** or until gelatin is dissolved. Stir in remaining orange juice and add lemon juice. Chill until slightly thickened. Fold in pineapple, carrot and cabbage; blend well. Pour into an 8x8x2-inch dish and refrigerate until set. Makes 6 to 8 servings.

CRANBERRY WALDORF SALAD

Freeze cranberries in fall for year-round use

- 1 (12-ounce) bag cranberries (3 cups)
- 2 cups hot water
- 1⅓ cups sugar
- 2 envelopes unflavored gelatin
- ¾ cup cold water
- 2 ribs celery, chopped
- 1 large apple, chopped (1¼ cups)
- ½ cup chopped walnuts

Chop cranberries in food processor or food grinder. Combine with hot water in a 2-quart casserole. Cover and **MW on High 5 to 5½ minutes,** stirring midway through cooking. Stir in sugar until dissolved.

Sprinkle gelatin over ¾ cup cold water and stir until dissolved. Add to cranberry mixture. Refrigerate until mixture begins to thicken. Add remaining ingredients; blend. Pour mixture into a 2-quart rectangular dish or large mold and refrigerate until set. Makes 12 servings.

Quick-Tip ➡ To yield more juice from citrus fruits (such as lemons and oranges), **MW on High 30 seconds per piece of fruit.**

Salads

GOLDEN COIN SALAD
In honor of our Las Vegas REVIEW JOURNAL

1 pound carrots, sliced into coins ⅛-inch thick
¼ cup water
1 medium onion, sliced ⅛-inch thick
1 green bell pepper, cut in narrow strips
⅔ cup vinegar
½ cup oil
1 (10¾-ounce) can condensed tomato soup
1 cup sugar
Salt and pepper to taste

Combine carrots and water in a 2-quart casserole. Cover with plastic wrap and **MW on High 8 to 10 minutes,** stirring once. Add onion and green pepper; re-cover and let stand while making sauce.

Combine vinegar, oil, soup, sugar, salt and pepper; pour over carrot mixture. Mix ingredients well. Refrigerate at least 24 hours for flavors to blend. Serve chilled. *Keeps 6 weeks.* Makes 6 to 8 servings.

KOREAN SPINACH SALAD
Jon Williamson's favorite

1 (10-ounce) package fresh spinach, washed and drained
1 (16-ounce) can bean sprouts, drained
3 hard-cooked eggs, sliced
8 slices bacon, cooked and crumbled

Tear spinach leaves and place in salad bowl. Add bean sprouts; toss. Garnish top with egg slices and sprinkle crumbled bacon on top of salad. Serve with Korean Dressing below. Makes 4 to 5 servings.

KOREAN DRESSING

¼ cup red wine vinegar
½ cup sugar
1 cup vegetable oil
⅓ cup catsup
1 teaspoon worcestershire sauce

In a 4-cup glass measure combine vinegar and sugar. **MW on High 1½ minutes,** or until sugar is dissolved. Add remaining ingredients; blend well. Refrigerate. Yields 2 cups.

Salads

CHALUPA SALAD
Mexican chef salad

- 1 pound ground beef
- ½ green bell pepper, chopped
- 1 cup chopped onion
- 1 tablespoon chili powder
- ½ teaspoon cumin
- 1 head iceberg lettuce, shredded
- 2 medium tomatoes, chopped
- 1 avocado, sliced
- 1½ cups shredded cheddar cheese (6 ounces)
- 12 chalupa shells
- Sour cream
- Taco sauce

Crumble beef into a hard-plastic colander set in a 1½-quart casserole. Sprinkle pepper and onion on top. **MW on High 6 to 7 minutes,** stirring midway through cooking. Discard grease and transfer beef mixture into same casserole. Add chili powder and cumin; mix well.

Combine lettuce, tomatoes and avocado in a salad bowl. Add warm meat mixture and top with cheese. Let stand a few minutes so that cheese will melt slightly. To serve, spoon onto a chalupa shell and top with a dollop of sour cream. Drizzle with taco sauce. Makes 12 chalupa salads.

HOT TURKEY SALAD

- 3 cups diced cooked turkey
- ¼ cup chopped onion
- 2 ribs celery, chopped
- 1 green bell pepper, chopped
- 1 cup mayonnaise
- ¼ teaspoon curry powder
- Salt and pepper to taste
- 1½ cups shredded cheddar cheese (6 ounces)
- Seasoned dry bread crumbs

If using food processor, chop turkey with steel blade; remove. Chop onion, celery and bell pepper together, using same blade.

Combine turkey, vegetables, mayonnaise, curry powder, salt and pepper in a 3-quart round casserole. **MW on 70% (Medium-high) 8 minutes.** Sprinkle cheese over top and dust with bread crumbs. **MW on 70% 4 to 5 minutes,** or until cheese has melted. Makes 4 to 6 servings.

Salads ←——————————————————

CHILLED CHICKEN PIE

1 envelope unflavored gelatin
1 cup cold water
1 (10¾-ounce) can condensed cream of chicken soup
4 dashes Tabasco
2 tablespoons lemon juice
¼ cup mayonnaise
1 (5-ounce) can boned chicken, diced
1 (5-ounce) can water chestnuts, drained and chopped
⅓ cup toasted almonds, sliced
1 (9-inch) pie shell, baked
1 (16-ounce) can jellied cranberry sauce

Sprinkle gelatin over ½ cup of the cold water in a 4-cup glass measure; stir. **MW on High 1 to 1½ minutes,** or until boiling. Stir to dissolve. Blend in remaining water, soup, Tabasco, lemon juice and mayonnaise. Chill mixture until slightly syrupy.

Fold in chicken, water chestnuts and almonds, and turn into pie shell. Refrigerate until set. To serve, garnish with slices of cranberry sauce cut into decorative shapes. Makes 6 servings.

SHRIMPLY SALAD

12 ounces frozen peeled and deveined shrimp
1 cup finely-chopped celery
½ cup chopped green bell pepper
½ cup slivered almonds
½ cup mayonnaise
1 tablespoon minced onion
2 teaspoons lemon juice
¼ teaspoon turmeric
¼ teaspoon salt
Pinch of cayenne pepper
1 cup grated potato chips
½ cup shredded cheddar cheese

NOTE: *If the frozen shrimp were cooked before packaging, defrost and drain; then add directly to other ingredients.*

For raw frozen shrimp, defrost and place in casserole with ½ cup water. **MW on High 3 minutes.** Drain.

Combine cooked shrimp with celery, bell pepper, almonds, mayonnaise, onion, lemon juice, turmeric, salt and pepper. Divide mixture into 4 sea shells or individual ramekins. **MW on High 3 to 4 minutes.** Sprinkle potato chips and cheese over shrimp mixture. **MW on 70% (Medium-high) 2 to 3 minutes.** Makes 4 servings.

Salad Dressings

AVOCADO DRESSING

- 1 ripe avocado
- 1 tablespoon lemon juice
- Several dashes Tabasco sauce
- ½ cup light cream
- ½ cup sour cream
- 1 clove garlic, minced
- 1 tablespoon grated onion
- ½ teaspoon salt
- Dash of cayenne pepper

Peel and mash avocado with lemon juice and Tabasco sauce. Add remaining ingredients and blend well; chill until serving time. Yields 1¼ cups.

HOT BACON DRESSING

- 4 slices bacon, diced
- ¼ cup finely chopped onion
- 2 tablespoons packed brown sugar
- 1 tablespoon flour
- ¼ teaspoon dry mustard
- 1 tablespoon lemon juice
- ¾ cup water
- ¼ cup mayonnaise

Place bacon in a 2-cup glass measure, cover with a paper towel and **MW on High 3 to 4 minutes,** or until nicely browned. Remove bacon and reserve. Add onion to bacon grease and **MW on High 2 to 3 minutes.** Stir in brown sugar, flour, dry mustard, lemon juice and water. **MW on High 1½ to 2 minutes,** or until thickened. Blend in mayonnaise until mixture is smooth. Add reserved bacon; stir. Yields 1¼ cups

ZIPPY CROUTONS

- 9 slices whole-wheat bread, cut into 1-inch cubes
- ½ cup margarine
- 1 tablespoon EACH: Parmesan cheese and Italian seasoning
- ¼ teaspoon EACH: marjoram, thyme and garlic powder
- ⅛ teaspoon curry powder

Place bread cubes in a 2-quart rectangular dish. **MW on High 6 minutes,** stirring twice during cooking. Place margarine in a 1-cup glass measure and **MW on High 1 minute,** or until melted. Drizzle margarine over bread cubes. Combine cheese and seasonings; sprinkle over top of bread cubes. Stirring after 2 minutes, then again after every minute, **MW on High 4 to 6 minutes,** or until croutons are toasted. Croutons will become crisp upon cooling. Yields 1½ quarts.

Salad Dressings

SUE'S COOKED SALAD DRESSING
Homemade without the jar

- ½ cup sweet pickle liquid
- ½ cup water
- ½ cup sugar
- 1 teaspoon dry mustard
- 1 tablespoon flour
- 3 eggs, beaten

Combine pickle liquid and water in a 4-cup glass measure. **MW on High 1½ minutes.** Blend sugar, dry mustard and flour together in a small bowl. Add beaten eggs to dry ingredients. Add egg mixture to hot pickle liquid; blend well. Whisking midway through cooking, **MW on 70% (Medium-high) 3 to 3½ minutes** or until thickened. Yields 2 cups.

Variation: POPPY SEED DRESSING. Add 1½ tablespoons poppy seeds and 1 tablespoon lime juice to above dressing.

PINEAPPLE-POPPY SEED DRESSING
Wonderful over fresh fruits

- 1 (6-ounce) can pineapple juice
- ⅔ cup sugar
- ½ teaspoon dry mustard
- 1 tablespoon white wine vinegar
- 1 tablespoon lemon juice
- 1 tablespoon minced onion
- 1 cup vegetable oil
- 1 tablespoon poppy seeds

Combine pineapple juice, sugar and mustard in a 2-cup glass measure. **MW on High 6 to 7 minutes,** or until mixture thickens slightly. *Note: Power level may need to be reduced to 70% (Medium-High) or lower if mixture begins to bubble out of glass measure.* Cool to room temperature.

Pour pineapple mixture into blender or food processor and add vinegar, lemon juice and onion. With blender operating, slowly add oil. Add poppy seeds; blend. Refrigerate. Yields 2 cups.

EGGS and QUICHE

Eggs

HAMBURGER-BOX OMELET

- 1 foam hamburger box from a fast-food restaurant
  Paprika
- ¼ cup chopped green bell pepper
- 1 green onion with top, thinly sliced
- ¼ cup diced cooked ham
- 2 eggs, well beaten
- 2 tablespoons milk
  Salt and pepper to taste
- 1 tablespoon shredded cheddar cheese

Spray inside of box with Pam (both bottom and lid). Sprinkle with paprika. Place bell pepper and onion in a 1-cup glass measure. Cover with plastic wrap and **MW on High 1 minute.** Add ham and re-cover; set aside.

Beat together eggs, milk, salt and pepper. Pour mixture equally into bottom and lid of prepared hamburger box. **MW on 50% (Medium) 2 to 2½ minutes,** or until almost set. Sprinkle reserved ham mixture over egg in bottom of box. Quickly close lid to flip top of omelet over. Lock lid shut and **MW on 50% 1 minute.** Open lid, sprinkle with cheese, and close lid for omelet to rest 1 minute before serving. Yields 1 omelet.

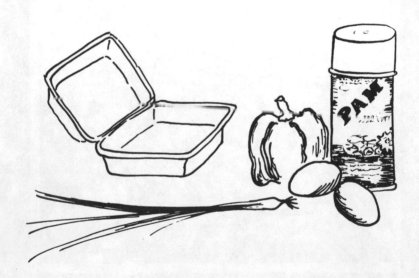

Eggs

BACON AND EGG BURRITO

- 2 slices bacon
- 1 egg
- 1 tablespoon milk
- 1 teaspoon chopped onion
- ¼ cup shredded cheese
- 1 flour tortilla
- Taco sauce

Arrange bacon on a folded paper towel and cover with another paper towel. Place on a microwave-safe dinner plate and **MW on High 2 minutes,** or to desired doneness. Combine egg, milk and onion in a 1-cup glass measure; beat well. **MW on High 50 to 60 seconds,** stirring midway through cooking.

Sprinkle cheese down center of tortilla. Top with egg, taco sauce and bacon. Fold tortilla sides over and place seam-side down on dinner plate. **MW on High 30 seconds,** or until heated through. *Garnish as desired.* Makes 1 serving.

BEV'S SUNSHINE BREAKFAST SANDWICHES

- 6 slices bacon
- 1 (3-ounce) package cream cheese with chives
- 2 tablespoons milk
- ⅓ cup shredded Swiss cheese
- 3 hard-cooked eggs, coarsely chopped
- 3 English muffins, split in half and toasted
- Paprika

Place a paper towel on a bacon or meat rack. Lay separated pieces of bacon on towel. Cover bacon with another paper towel. Rotating rack once or twice, **MW on High 4½ to 4¾ minutes,** or until bacon is done. Crumble bacon and reserve.

Place cream cheese in a 4-cup glass measure and **MW on 50% (Medium) 45 seconds,** or until softened. Use a whisk to blend in milk. Stir in cheese, eggs and reserved bacon. To serve, divide mixture among toasted muffin halves. Sprinkle with paprika. Place sandwiches on paper toweling. For each muffin half, **MW on 70% (Medium-high) 20 to 25 seconds.** If mixture has been refrigerated, add 15 seconds. Yields 1½ cups.

ONION-TOMATO CHEESE PIE

A favorite of Ann's mom, Ruth Beddow

   2 large onions, thinly sliced
   2 tablespoons margarine
   1 (10-inch) pie shell, baked
   2 tablespoons flour
2½ cups shredded Swiss or gruyere cheese
   2 firm tomatoes, sliced
   1 (5⅓-ounce) can evaporated milk
   2 eggs
   1 teaspoon Italian seasoning

Combine onions and margarine in a 2-quart batter bowl. Cover and **MW on High 5½ to 6½ minutes.** Remove onions with a slotted spoon and place in bottom of pie shell. Stir flour into cheese and sprinkle over top of onions. Place tomatoes on top of cheese.

To liquid remaining in batter bowl, add milk, eggs and Italian seasoning; beat well. Pour milk mixture over ingredients in pie shell. Rotating midway through cooking, **MW on 70% (Medium-high) 10 to 12 minutes,** or until center is almost set. Let stand 10 minutes before cutting. Makes 6 to 8 servings.

SHRIMP AND AVOCADO QUICHE

In honor of our SACRAMENTO BEE

- 1 cup shredded mozzarella cheese (4 ounces)
- 1 (9-inch) pie shell, baked
- 1 ripe avocado, sliced
- 2 teaspoons lemon juice
- 1 cup cooked or canned shrimp
- 3 eggs
- ⅔ cup light cream
- 2 tablespoons chili sauce
- ⅛ teaspoon cayenne pepper
- ⅛ teaspoon basil
- 1 green onion with top, thinly sliced

Place half of cheese in bottom of pie shell. Toss avocado with lemon juice to prevent darkening. Distribute over cheese. Add shrimp and remaining cheese. In a small bowl, beat together eggs, cream, chili sauce, cayenne pepper and basil. Pour over mixture in pie shell. Sprinkle green onions over top. **MW on 70% (Medium-high) 10 to 12 minutes,** rotating dish once midway through cooking. Let stand 10 minutes before cutting. Makes 6 servings.

SPINACH QUICHE

- 1 (10-ounce) package frozen chopped spinach
- 1 cup light cream or half and half
- 3 eggs
- 2 green onions with tops, thinly sliced
- ½ teaspoon salt
- ¼ teaspoon pepper
- ⅛ teaspoon nutmeg
- 1 cup shredded Swiss cheese (4 ounces)
- 1 (9-inch) pie shell, baked

Stand box of unopened spinach in a 4-cup glass measure. **MW on High 5 minutes.** Squeeze remaining water out of spinach and discard liquid. Set spinach aside.

In same glass measure, combine cream and eggs; beat well. Add onions, salt, pepper and nutmeg. Fold in cheese and reserved spinach. Pour mixture into pie shell. Sprinkle with paprika for garnish. **MW on 70% (Medium-high) 10 to 12 minutes,** rotating dish once midway through cooking. Let stand 10 minutes before cutting. Makes 6 servings.

IMPOSSIBLE QUICHES

Make the impossible possible—in your microwave oven! Impossible Pies are a more recent variation of quiche, which is a custard type of pie that is usually served as a luncheon or supper dish. The Impossible Pie concept was developed by General Mills, and their conventional recipes use Bisquick.

As the quiche, or Impossible Pie, bakes, it forms its own crust. A conventionally-baked Impossible Pie browns from the heat in the oven. The microwave variation of the Impossible Pie is the same color after microwaving as it was before. Therefore, we use garnishes on the top, such as French fried onion rings, olives, chopped tomatoes or paprika, to add color.

Impossible Pies, like quiches, contain an ample amount of cheese, milk and eggs. The delicate nature of these dairy products requires a lower power setting to yield a smooth uniform product. Therefore, we prefer to microwave Impossible Pies and quiches on 70% (Medium-high). Rotating during microwaving and standing time afterwards are most important for best results.

IMPOSSIBLE CHEESY TUNA PIE

- ½ cup chopped celery
- ½ cup chopped onion
- 2 cups shredded cheddar cheese (8 ounces)
- 1 (6½-ounce) can tuna, drained
- 1¼ cups milk
- 3 eggs
- ½ cup buttermilk baking mix
- ⅛ to ¼ teaspoon ground rosemary or thyme
- Pepper to taste
- Paprika
- 1 (2¼-ounce) can sliced ripe olives, drained

Combine celery and onion in a 1 cup glass measure. Cover with plastic wrap and **MW on High 3 minutes.** Place mixture in a 10-inch glass pie plate which has been sprayed with Pam. Sprinkle 1 cup cheese over top of vegetables. Distribute tuna over top of cheese. Sprinkle remaining cheese over tuna.

Combine milk, eggs, baking mix, rosemary and pepper in blender or food processor. Process until blended. Pour over cheese and tuna mixture. Sprinkle paprika over top. Distribute olives over top. Rotating dish every 3 minutes, **MW on 70% (Medium-high) 14 to 16 minutes,** or until center is set. Let stand 5 to 7 minutes before cutting. Makes 6 servings.

Quiche

IMPOSSIBLE QUICHE LORRAINE

- ½ pound bacon, cooked and crumbled, reserving drippings
- 1¾ cups milk
- 4 eggs
- 2 cups shredded Swiss cheese (8 ounces)
- ½ cup buttermilk baking mix
- 1 tablespoon snipped chives
- Salt and pepper to taste
- ½ (3-ounce) can French fried onion rings

Place one tablespoon reserved bacon drippings in a 10-inch glass pie plate. Tilt to coat utensil. Sprinkle bacon in pie plate.

Combine 2 tablespoons bacon drippings, milk, eggs, cheese, baking mix, chives, salt and pepper in blender or food processor. Process until blended. Pour over bacon. Sprinkle onion rings on top. Rotating dish every 3 minutes, **MW on 70% (Medium-high) 13 to 15 minutes,** or until center is set. Let stand 5 to 7 minutes before cutting. Makes 6 servings.

GARDEN PATCH IMPOSSIBLE PIE

- 2 cups thinly sliced zucchini
- 4 green onions with tops, thinly sliced
- 4 ounces fresh mushrooms, sliced
- 1 cup milk
- 3 eggs
- 1½ cups shredded Jarlsberg cheese (6 ounces)
- ½ cup buttermilk baking mix
- 1 tablespoon parsley flakes
- ½ teaspoon basil
- ½ teaspoon oregano
- 2 medium tomatoes, sliced

Combine zucchini, onion and mushrooms in a 2-quart batter bowl. Cover with plastic wrap and **MW on High 4 to 4½ minutes.** Use slotted spoon to transfer cooked vegetables to a 10-inch glass pie plate which has been sprayed with Pam.

Combine remaining vegetable liquid, milk, eggs, cheese, baking mix and seasonings in blender or food processor. Process until blended. Pour over vegetables. Rotating dish every 3 minutes, **MW on 70% (Medium-high) 12 to 13 minutes.** Arrange tomato slices on top. **MW on 70% 3 minutes,** or until center is set. Let stand 5 to 7 minutes before cutting. Makes 6 servings.

Quiche ←

TACO GRANDE PIE

- 1 pound ground beef
- ½ cup chopped onion
- 1 to 3 teaspoons chili powder
- 1 (4-ounce) can chopped green chilies, drained
- 1½ cups milk
- 3 eggs
- 2 cups shredded cheddar cheese (8 ounces)
- ½ cup buttermilk baking mix
- 1 cup sour cream
- 1½ cups shredded lettuce
- 2 medium tomatoes, diced

Crumble ground beef into a hard-plastic colander set in a 1½-quart round casserole. Sprinkle onion on top. **MW on High 6 minutes,** stirring midway through cooking. Discard grease and transfer meat mixture into same casserole. Blend in chili powder and green chilies. Distribute meat mixture evenly in a 10-inch glass pie plate which has been sprayed with Pam.

Combine milk, eggs, 1½ cups cheese and baking mix in blender or food processor. Process until blended. Pour over meat mixture. Rotating dish every 3 minutes, **MW on 70% (Medium-high) 14 to 16 minutes,** or until center is set. Let stand 5 to 7 minutes before garnishing and cutting.

To garnish, spread sour cream over top of pie. Distribute lettuce around outer edge, then place tomatoes next to lettuce. Sprinkle remaining cheese in center of pie. Makes 6 servings.

Dear Microscope,
   Foods I cook in the microwave oven always seem dried out. What is the cause of this?
Dear Microcook,
   Overcooking is the cause of dry, tough or rubbery foods. Most people overcook foods because they do not allow for standing time or carryover cooking time. When microwaves enter foods, they cause the molecules to vibrate very rapidly. This creates friction which generates heat for cooking. When the microwave oven turns off, or if you take the food out of the microwave oven, the food molecules continue to vibrate for one-third to one-half the cooking time. During the time, the food continues to cook. If you microwave foods too long, this standing time results in overcooked food. Learn to trust a reliable cookbook written for your wattage of oven, or weigh food to determine correct cooking time.

PASTA & CHEESE MAIN DISHES

SPAGHETTI AND PASTA

Are there strings attached to microwave cooking? We hope so, if you're microwaving spaghetti! In Italian, "spaghetti" means "strings."

Whether you're cooking the pasta, the sauce, or both, spaghetti is super in a microwave oven. If you're in a hurry for dinner, the best use of your time is to cook the pasta conventionally at the same time you're microwaving the sauce. That way, you can eat in about 20 minutes.

You could, instead, microwave the pasta while you're assembling the sauce ingredients, fixing a salad, or spreading the garlic bread, see page 61.

Microwaving produces cooked pasta which native Italians would applaud! It is "al dente"–soft, but slightly resistant to the bite. The method used to cook pasta is the same via microwave as it is on your stove.

When you cook pasta, you should cook a double batch. **One pound of spaghetti weighs four pounds after cooking, and serves 12.** After cooking pasta and draining it in a colander, toss it with one or two tablespoons of olive oil, butter or margarine to prevent strands from sticking together.

Freeze the extra pasta for quick rejuvenation in the microwave oven some other busy day! Place cooked pasta in freezer-weight zip-locking plastic bags. Spread it out in an even, flat layer to speed defrosting. To reheat, unzip the bag part way and **MW on High about 2 minutes per cup,** redistributing the pasta as it defrosts to aid uniform heating.

Spaghetti sauce variations are limitless. If you are a one or two-person household, don't bypass recipes just because they yield six servings. Freeze the leftover sauce in portions and treat yourself to effortless Italian dinners on future nights.

BASIC PASTA

- 6 cups hottest tap water
- 1 tablespoon oil
- 1 teaspoon salt
- 8 ounces pasta

Place water in a 4-quart simmer pot or casserole. Cover and **MW on High 8 minutes,** or until water boils. Add oil, salt and pasta; stir to prevent strands from sticking together. Cover. Stirring once midway through cooking, **MW on High 6 to 8 minutes,** or until pasta tests "al dente." Drain and rinse in colander. Makes 6 servings.

NOTE: *To prevent pasta from sticking together after cooking, return to pot and toss with olive oil or margarine.*

Quick-Tip ➡ If you hate cooking lasagne noodles, you'll love our recipes using uncooked noodles! See pages 90 and 91 in our *Microwave Know-How* book, *MicroScope Savoir Faire.* (See coupon in back for ordering information.)

EASY MEATBALL SPAGHETTI SAUCE

- 1 (48-ounce) jar prepared spaghetti sauce
- ⅓ cup red wine (optional, but best)
- 1 pound ground beef
- ½ cup Italian seasoned dry bread crumbs
- 2 eggs

Reserve ¼ cup sauce. Place remaining sauce and wine in a 2-quart casserole. Cover and **MW on High 8 minutes.** Meanwhile, make meatballs. Place two paper towels in bottom of a 2-quart rectangular dish. In a mixing bowl, combine beef, reserved spaghetti sauce, crumbs and eggs. Form into (1-inch) balls and place on paper towels in dish. Cover with wax paper and **MW on High 3 minutes.** Use a spoon to move meatballs from the outside of the dish toward the center, and vice versa. Re-cover and **MW on High 3 to 4 minutes.** Add cooked meatballs to heated spaghetti sauce. **MW on High** to desired serving temperature. Serve over cooked pasta. Makes 6 servings.

PASTA PRIMAVERA

- 2 cups fresh broccoli flowerets
- 6 fresh asparagus spears, cut into pieces
- 2 cloves garlic, minced
- ¼ cup olive oil
- 1 teaspoon basil
- ¼ pound fresh mushrooms, sliced
- 1 zucchini squash, sliced ¼" thick
- 2 medium tomatoes, chopped coarsely
- 1 (6-ounce) package frozen snow peas, defrosted
- 1 (10¾-ounce) can tomato puree
- ¼ cup white wine
- ¼ cup water

Combine broccoli, asparagus, garlic, olive oil and basil in a 2-quart casserole. Cover and **MW on High 4 minutes.** Add mushrooms, squash and tomatoes. Re-cover and **MW on High 4 minutes.** Add snow peas, tomato puree, wine and water. Re-cover and **MW on High 6 minutes,** or until heated through. Serve over cooked pasta. *Garnish with grated parmesan cheese, if desired.* Makes 6 servings.

NOTE: *This spaghetti sauce does not freeze as well as the others because of its fresh vegetables, whose quality lessens during freezing.*

Pasta

LIEUTENANT KELLY'S BALBOA ISLAND SPAGHETTI

- 1 pound ground beef
- 1 medium onion, chopped
- 1 (28-ounce) can tomatoes and liquid
- 1 (8-ounce) can tomato sauce
- 1 (6-ounce) can tomato paste
- 1 (4-ounce) can mushroom pieces, drained
- ⅓ cup red wine
- ¼ cup sliced ripe olives
- 2 bay leaves
- 2 teaspoons basil
- 1 teaspoon oregano
- ½ teaspoon instant minced garlic; or 1 clove fresh garlic, minced
- Salt and coarsely-ground black pepper to taste

Crumble ground beef into a hard-plastic colander set in a 3 to 4-quart casserole. Sprinkle onion on top. **MW on High 6 to 7 minutes,** stirring midway through cooking. Discard grease and transfer meat mixture into same casserole. Add tomatoes, including liquid, and mash coarsely with a fork. Add tomato sauce, tomato paste, mushrooms, wine, olives, bay leaves, basil, oregano and garlic. Cover and **MW on High 15 minutes,** stirring once midway through cooking. (If you have more time, continue to **MW on 30% (Medium-low) 15 minutes,** for more blending of flavors.) Add salt and pepper to taste. Serve over cooked pasta. Makes 6 servings.

> **Pasta**

1-MINUTE GARLIC BREAD
You can't get much faster!

- ½ cup butter or margarine
- 3 cloves garlic
- ½ teaspoon oregano
- ½ teaspoon basil
- 1 pound loaf Italian bread sliced ¾-inch thick
- Grated parmesan cheese

Place butter in a small glass mixing bowl and **MW on 30% (Medium-low) 1 minute.** Using a garlic press, squeeze garlic into butter, and add residue from pulp. Blend in oregano and basil. Spread mixture on bread slices and sprinkle with cheese. Place slices together to form a loaf.

Line a microwave-safe wicker basket or utensil with six paper napkins or paper towels. Place bread in utensil and cover with paper napkins. **MW on High 1 minute** immediately before serving. Makes 20 slices.

SPAGHETTI PIE

- 6 ounces spaghetti
- 2 tablespoons margarine
- ¼ cup Parmesan cheese
- 1 egg, beaten
- 1 pound ground beef
- ½ cup chopped onion
- ¼ cup chopped green bell pepper
- 1 (6-ounce) can tomato paste
- 1 (8-ounce) can tomato sauce
- 1 clove garlic, minced
- ½ teaspoon oregano
- ½ teaspoon basil
- 1 cup small-curd cottage cheese
- Fresh parsley and Parmesan cheese for garnish

Cook spaghetti according to package directions; drain. Stir margarine into hot spaghetti. Blend in Parmesan cheese and egg. In a buttered 10-inch pie plate, shape spaghetti mixture into a crust. Set aside.

Crumble ground beef into a hard-plastic colander set in a 2-quart casserole. Sprinkle onion and green pepper on top.

MW on High 6 to 7 minutes, stirring once midway through cooking. Discard grease and place meat mixture in same casserole. Add tomato paste, tomato sauce, garlic, oregano and basil; blend well. **MW on High 2 to 3 minutes.**

Spread cottage cheese over spaghetti crust. Top with meat mixture. **MW on 70% (Medium-high) 6 to 7 minutes,** rotating once. Garnish with parsley and Parmesan cheese. Makes 6 servings.

Pasta

SPAGHETTI-STUFFED PEPPERS

- 4 large green bell peppers
- 3 cups cooked spaghetti
- 1 (15½-ounce) jar prepared spaghetti sauce or 1½ cups homemade spaghetti sauce
- 1 cup shredded mozzarella cheese (4 ounces)

Prepare peppers by cutting off just enough of the top to remove seeds; wash. Place peppers on a microwave-safe tray. Cover with plastic wrap and **MW on High 6 to 7 minutes,** rotating tray once. Let stand, covered, while preparing filling.

Place spaghetti in a 1½-quart round casserole. If refrigerated, cover and **MW on High 3 minutes.** Reserve ¼ cup sauce; blend remaining sauce and ¾ cup cheese with spaghetti. Fill peppers. Cover with plastic wrap and **MW on 70% (Medium-high) 4 to 5 minutes,** rotating tray once. Drizzle remaining sauce over top of peppers; sprinkle remaining cheese on top of sauce. **MW on 70% 1 to 2 minutes.** Let stand 3 minutes. Makes 4 servings.

SPINACH FETTUCCINE

- 8 cups hottest tap water
- 1 tablespoon oil
- 1 teaspoon salt
- 16 ounces fresh spinach fettuccine
- ½ cup butter
- 1 clove garlic, minced
- 1 teaspoon Italian seasoning

Place water in a 4-quart simmer pot or casserole. Cover and **MW on High 8 minutes,** or until water boils. Add oil, salt and pasta; stir to prevent strands from sticking. Cover. Stirring once midway through cooking, **MW on High 5 to 6 minutes,** or until pasta tests "al dente." Drain.

Combine butter, garlic and Italian seasoning in a 1-cup glass measure. **MW on High 1 minute,** or until butter is melted. Toss in cooked pasta. Makes 6 servings.

Pasta

MACARONI AND HAM SALAD

- 6 cups hottest tap water
- 1 tablespoon oil
- 1 (10-ounce) package shell macaroni
- 1 cup diced cooked ham
- 2 ribs celery, chopped
- 3 green onions with tops, thinly sliced
- 3 hard-cooked eggs, chopped
- 1 large tomato, diced
- Mayonnaise or Sue's Cooked Salad Dressing, page 48

Place water in a 4-quart simmer pot or casserole. Cover and **MW on High 6 minutes,** or until water boils. Add oil and macaroni shells; stir to prevent shells from sticking. Cover. Stirring once midway through cooking, **MW on High 7 minutes,** or until tender. Drain. Add remaining ingredients to macaroni shells, using just enough mayonnaise to moisten. Serve on crisp lettuce. Makes 5 to 6 servings.

INSIDE-OUT "MANICOTTI"

- 1 (16-ounce) jar prepared extra-thick spaghetti sauce
- ¼ cup dry red wine
- 8 ounces ziti or mostaccioli pasta, cooked
- 1 (16-ounce) carton large-curd cottage cheese
- 2 cups shredded mozzarella cheese (8 ounces)
- 1 tablespoon parsley flakes
- ½ teaspoon basil
- ½ teaspoon garlic powder
- 1 egg, slightly beaten
- Italian-flavored dry bread crumbs

Combine spaghetti sauce and wine. Pour half into a 2-quart casserole. Distribute half of the cooked pasta over sauce. Combine cheeses, spices and eggs. Spread half of the cheese mixture over pasta. Repeat layers with remaining ingredients. Garnish top of casserole with bread crumbs. Cover. Rotating midway through cooking, **MW on 70% (Medium-high) 15 to 17 minutes,** or until heated through. Makes 6 servings.

MICROWAVING MANICOTTI

This section is for cooks who hate boiling and stuffing manicotti noodles—and also for cooks who have never tried making this Italian specialty. We have developed an easy way: stuff the manicotti uncooked and then microwave them in a sauce.

The Italian word "cannelloni," a pasta similar to manicotti, translates into English as "big pipes," which is descriptive of these two. They are stuffed, usually with a cheese filling but also with meat or other ingredients, and then baked in a sauce. Pasta from southern Italy is most often manufactured in tubular form. It is frequently made without eggs and bought at the store in a dried form that makes it possible to keep it for long periods before cooking.

In making Manicotti conventionally, the noodles must be cooked first in boiling water; then drained and cooled. It is very difficult to fill these limp, slippery tubes, and very often, they split or break.

To microwave our manicotti, we stuff the pasta *before* cooking them, which is infinitely easier. It also eliminates the added time and trouble of cooking them in boiling water. A spatula, spoon or knife is helpful for stuffing, but in all honesty, there's sometimes no substitute for fingers!

The secret of successful manicotti microwaving is to heat the sauce, which aids the softening of the dried pasta. Pour half of the heated sauce into the casserole before adding the filled manicotti. Top with the remaining heated sauce.

Cover the casserole with plastic wrap to hold in the steam. After the first stage of microwaving, turn each manicotti over and spoon some sauce over the exposed side. Finish microwaving on reduced power so that the noodles can absorb liquid. Remember to let the manicotti stand for at least fifteen minutes after microwaving to allow for carry-over cooking.

Pasta

SAM'S CHEESE MANICOTTI

For our microwave friend, Sam Beneski

- ½ pound ground beef
- 1 (1½-ounce) packet spaghetti sauce mix
- 1 (6-ounce) can tomato paste
- 2½ cups water
- 8 ounces cottage cheese (1 cup)
- 1 cup shredded mozzarella cheese (4 ounces)
- ¼ cup grated parmesan cheese
- 1 egg, well beaten
- 1 tablespoon parsley
- ¼ teaspoon pepper
- 8 uncooked manicotti

Crumble ground beef into a 4-cup glass measure. **MW on High 3 minutes,** breaking up meat midway through cooking. Pour off grease. Stir in sauce mix, then tomato paste and water. Cover with plastic wrap and **MW on High 10 minutes.**

To make cheese filling, combine cheeses, egg, parsley and pepper. Fill uncooked manicotti, using a small spatula, spoon or knife.

Pour half of heated sauce into a 2-quart rectangular dish. Arrange filled manicotti in dish and pour remainder of sauce over them. Cover with plastic wrap. **MW on High 10 minutes.**

Using tongs or two forks, turn each manicotti over. Spoon some sauce over each. Re-cover and **MW on 70% (Medium-high) 15 to 17 minutes.** Let stand at least 15 minutes before serving. Makes 4 servings.

SAUSAGE MANICOTTI FLORENTINE

- 1 (10-ounce) package frozen chopped spinach, cooked and drained
- 8 ounces bulk sausage, cooked and drained
- ½ cup sour cream
- 1 teaspoon oregano
- ½ teaspoon basil
- 8 uncooked manicotti
- 1 (16-ounce) jar prepared extra-thick spaghetti sauce
- 1½ cups water

Combine spinach, sausage, sour cream, oregano and basil. Fill uncooked manicotti. Combine spaghetti sauce and water in a 4-cup glass measure. Cover with plastic wrap and **MW on High 6 minutes,** or until hot. Microwave according to last two paragraphs in recipe above. Makes 4 servings.

Pasta

MEXICAN MANICOTTI

1 (10-ounce) can mild enchilada sauce
1½ cups water
½ pound lean ground beef
1 cup canned refried beans
1 teaspoon oregano
½ teaspoon ground cumin
8 uncooked manicotti
1 cup sour cream
¼ cup sliced green onions with tops
¼ cup sliced pitted ripe olives
½ cup shredded cheddar cheese

Combine enchilada sauce and water in a 4-cup glass measure. Cover with plastic wrap and **MW on High 6 minutes.**

To make filling for manicotti, combine uncooked ground beef, refried beans, oregano and cumin. Fill uncooked manicotti, using a small spatula, spoon or knife.

Pour half of heated sauce into a 2-quart rectangular dish. Arrange filled manicotti in dish and pour remainder of sauce over them. Cover with plastic wrap. **MW on High 10 minutes,** rotating dish midway through cooking.

Using tongs or two forks, turn each manicotti over. Spoon some sauce over each. Re-cover, and rotating dish midway through cooking, **MW on 70% (Medium-high) 15 to 17 minutes.** Let stand 15 minutes. Immediately before serving, spoon sour cream down center of manicotti. Sprinkle with green onions, olives and cheese. Do not cover. **MW on 70% 2 to 3 minutes,** or until cheese melts. Makes 4 servings.

Quick-Tip ➡ Leftover fondue can be refrigerated and reheated. **MW on 70% (Medium-high) until heated through.** Stirring the fondue at least once during reheating will speed the melting and heating of the fondue.

> Fondue

CATHY'S CHEDDAR FONDUE

A favorite of one of Ann's two daughters

- 4 tablespoons margarine
- 4 tablespoons flour
- ½ teaspoon dry mustard
- ¼ teaspoon paprika
- 1 teaspoon worcestershire sauce
- 2 cups milk
- 2 cups shredded cheddar cheese (8 ounces)

Place margarine in a 1½-quart round casserole. **MW on High 45 seconds,** or until melted. Stir in flour, mustard and paprika. Blend worcestershire into milk and add to flour mixture. Whisking mixture midway through cooking, **MW on High 4½ to 5½ minutes,** or until thickened. Stir in cheese. **MW on 70% (Medium-high) 2 to 3 minutes,** or until cheese is melted. *Serve with cubes of French bread or ½-inch pieces of cooked sausage.* Yields 3¼ cups.

PIZZA FONDUE

Our families gave this a "10" for Saturday night supper!

- ½ pound bulk sausage or ground beef
- ½ cup chopped onion
- 1 (14-ounce) jar pizza sauce
- 1 (8-ounce) can tomato sauce
- 1 teaspoon EACH: basil, oregano, fennel seed and garlic powder
- 1½ cups shredded mozzarella cheese (6 ounces)

Crumble sausage into a hard-plastic colander set in a 1½-quart round casserole. Sprinkle onion on top.

MW on High 4 to 4½ minutes, stirring midway through cooking. Discard grease and transfer meat mixture into same casserole. Add pizza sauce, tomato sauce, basil, oregano, fennel seed and garlic powder.

MW on High 3 minutes, stirring once midway through cooking. Blend in cheese and **MW on 70% (Medium-high) 2 to 3 minutes,** or until cheese is melted. *Serve with toasted cubes of French bread.* Yields 1½ quarts.

Cheese Casseroles

AMERICAN MAH-JONG

- 1 cup chopped green bell pepper
- ½ cup chopped onion
- 3½ cups cooked rice
- 1 (28-ounce) can tomatoes, drain and reserve juice
- ¾ cup chopped spanish salad olives
- ¼ teaspoon pepper
- ¼ teaspoon turmeric
- 1 (6-ounce) roll processed cheese, cubed

Combine green pepper and onion in a 2-quart casserole. Cover and **MW on High 3 minutes.** Add rice. Chop tomatoes and add to rice mixture along with olives, pepper and turmeric; blend well. Stir in cheese and add ½ cup reserved juice from tomatoes. (Additional tomato juice can be added if you desire.) Stirring midway through cooking, **MW on 70% (Medium-high) 9 to 11 minutes,** or until heated through. Let stand 5 minutes before serving. Makes 6 servings.

NACHO FIESTA CASSEROLE

- 1 large onion, chopped
- 2 tablespoons margarine
- 2 eggs, slightly beaten
- 1 cup small-curd creamed cottage cheese
- ¼ teaspoon ground cumin
-   Dash pepper
- 1 (7-ounce) package nacho cheese tortilla chips
- 1 (4-ounce) can chopped green chilies (do not drain)
- 3 medium tomatoes, chopped
- 2 cups shredded Monterey Jack cheese (8 ounces)
- 1 cup sour cream
- 1 cup shredded sharp cheddar cheese (4 ounces)

Place onion and margarine in a 4-cup glass measure. Cover and **MW on High 2 to 2½ minutes,** stirring midway through cooking. Blend in eggs, cottage cheese, cumin and pepper.

Crush enough chips to make ½ cup. Set aside. Place half of remaining chips evenly in greased 2-quart rectangular glass dish. Layer with half of egg-cheese mixture, half of chilies, half of tomatoes, and half of Monterey Jack cheese. Repeat layers and cover. **MW on High 8 minutes,** rotating dish every 2 minutes. Let stand 10 minutes. Spread sour cream over casserole. Sprinkle with cheddar cheese and reserved crushed chips. **MW, uncovered, on High 3 minutes.** Let stand 10 minutes before serving. Makes 8 servings.

MEAT
POULTRY
FISH

THE ALL-AMERICAN HAMBURGER

In America more meat is used for hamburgers than any other purpose. Learn the many ways a microwave can be your helper with hamburgers—from defrosting to reheating.

We'll start at the store. For economy, regular ground beef is a better buy than extra-lean ground beef. According to Dr. Russell Cross of the United States Department of Agriculture, "There is practically no difference in cooked hamburgers whether made from extra-lean or regular ground beef, except that hamburgers made from regular beef are juicer and a bit tastier." They contain almost identical amounts of protein. As hamburgers are microwaved, regular ground beef will lose fat while extra-lean ground beef will lose water.

When you get home from the store, spend a few minutes preparing your ground beef for the freezer. Meat patties will defrost much faster than a block of beef, especially if they can be separated. From each pound of ground beef, shape four patties on squares of wax paper, and top with more squares of wax paper. Stack according to the number you need for a meal, and wrap for the freezer.

HOW TO DEFROST HAMBURGERS

Remove freezer wrapping and separate patties by wedging a dinner knife between the layers of wax paper. Place patties on a microwave meat or bacon rack. If you will be cooking the hamburgers immediately, you can **defrost patties on High power, allowing 45 seconds to 1 minute per pattie.** Rotate rack midway through defrosting. Let stand to complete defrosting.

HOW TO DEFROST HAMBURGER BUNS

Remove the metal twist-tie from the plastic bag and slip a double layer of paper toweling between the buns and the plastic bag, both top and bottom. Place bag on a microwave meat or bacon rack. For a package of **8 hamburger buns, MW on 30% (Medium-low) 2½ to 3 minutes.** Rotate package top to bottom midway through defrosting. Redistribute each layer of 4 buns so that the least defrosted areas are moved to the outside. Separate buns and let stand to finish defrosting. If you microwave bread products too long, they will be tough and chewy.

MICROWAVING HAMBURGERS

Hamburgers can be microwaved on a microwave meat or bacon rack. If using this method, you will need to apply "cosmetics" so that the patties will have an appealing color. Before microwaving, brush patties with Kitchen Bouquet or microwave browning sauce, or sprinkle moistened patties with Micro Shake. To microwave 1 pound (4 patties) on a meat or bacon rack, leave uncovered or cover lightly with wax paper. To minimize shrinkage, **MW on 70% (Medium-high) 5 to 6 minutes,** turning each pattie halfway round after 3 minutes.

BROWNING DISH HAMBURGERS

To microwave hamburgers using a browning dish, preheat empty dish according to manufacturer's instructions. Place **4 patties** in preheated browning dish, pressing meat with spatula to contact hot surface evenly. **MW on High 1½ minutes.** Turn over and **MW on High 1½ to 2½ minutes,** or until almost done. Let stand 2 minutes.

REHEATING HAMBURGERS

Another use for your microwave is reheating already-cooked hamburgers. We do not recommend reheating an assembled hamburger, as the meat juices will make the bun soggy. **To reheat 4 hamburger buns** without meat, wrap in paper towels and place on a rack. **MW 25 to 30 seconds on High. To reheat hamburger patties, MW on 70% (Medium-high) 30 seconds per pattie.**

If your preference is to grill hamburgers outdoors, undercook an extra batch and **freeze individual hamburgers** between squares of wax paper. When you need a meal in a hurry, separate the number of patties you need and **MW on 70% (Medium-high)** until defrosted and heated through. They will taste as charcoaled as the day you grilled them.

LUBBOCK BACON BURGERS

In honor of our Lubbock Avalanche Journal

- 4 hamburger patties (1 pound)
- 4 slices bacon
- 3 medium onions, thinly sliced
- Bottled steak sauce
- 1 cup shredded Longhorn cheese (4 ounces)
- 4 hamburger buns

Cook hamburger patties as desired and set aside. Place bacon in a 10-inch square casserole. Cover and **MW on High 5 minutes.** Remove bacon and drain on paper towel. Add onions to bacon drippings. Cover and **MW on High 8 minutes,** or until onions are transparent. Remove onions and place patties in casserole. Drizzle with steak sauce, distribute onions then cheese over meat. Crumble bacon and garnish patties. **MW on 70% (Medium-high)** until cheese melts. Place on buns. Makes 4 servings.

Hamburgers

KATHLEEN'S BRONTOBURGERS

If you can't get dinosaur meat, substitute ground beef!

- **1 pound ground dinosaur meat or ground beef**
- **1 pterodactyl egg or hen's egg**
- **1 chopped fern or green onion**
- **½ cup crushed bedrock or uncooked oats**
- **¼ cup lukewarm lava or barbecue sauce**

Mix dinosaur meat, pterodactyl egg, chopped fern, crushed bedrock and lukewarm lava. Form into four patties. Place on meat or bacon rack. Cover with wax paper. **MW on High 5 to 6 minutes,** turning each pattie halfway around after 3 minutes. To serve, place patties on hamburger buns and drizzle with lukewarm lava. Add other condiments as desired. Makes 4 servings.

NOTE: *These are also great grilled over the barbecue.*

ROQUEFORT BURGER TOPPING

Also good over steaks

- **2 teaspoons oil**
- **1 clove garlic, minced**
- **1 teaspoon brandy**
- **½ cup crumbled Roquefort cheese**

Combine ingredients in a small bowl and spoon over hamburger patties 1 minute before they are finished microwaving. Makes topping for 4 patties.

➡ *Hamburgers*

CONTINENTAL HAMBURGERS

4 hamburger patties (1 pound)
8 ounces fresh mushrooms, sliced
1 tablespoon olive oil
4 slices Swiss cheese
½ teaspoon dill weed
8 slices pumpernickel bread
Dijon-style mustard

Cook hamburger patties as desired and set aside. Place mushrooms and olive oil in a 4-cup glass measure. Cover with plastic wrap and **MW on High 3 to 4 minutes.** Place patties on a microwave-safe plate. Top with cheese and cooked mushrooms. Dust with dill weed. **MW on 70% (Medium-high)** until heated through. Spread bread with mustard and assemble hamburgers. Makes 4 servings.

DOUBLE-UP HAMBURGERS

1 medium onion, finely sliced and separated into rings
1 tablespoon margarine
1½ pounds ground beef
¼ cup fine bread crumbs
¼ cup barbecue sauce
½ cup shredded Monterey Jack cheese
Kitchen Bouquet or Micro Shake

Combine onion and margarine in a 2-cup glass measure. **MW on High 3 to 3½ minutes,** or until tender. Set aside.

Combine ground beef, bread crumbs and barbecue sauce. Form 8 flat patties. Divide onion among 4 patties and sprinkle cheese over top of onion. Top each filled patty with a plain patty and crimp edges together to seal. Brush Kitchen Bouquet or sprinkle Micro Shake on hamburgers before placing on meat or bacon rack. Cover with wax paper and **MW on High 8 to 9 minutes,** turning each pattie halfway around after 3 minutes. Let stand 3 minutes. Makes 4 servings.

Quick-Tip ➡ For the calorie conscious, hamburgers can be enjoyed with only a bottom bun or meat and trimmings alone.

Hamburgers ←

HAMBURGERS TERIYAKI

4 hamburger patties (1 pound)
Teriyaki marinade
2 center slices of a large onion
8 strips green bell pepper
4 slices pineapple, drained
Sweet-and-sour sauce (optional)

Place patties in a 10-inch square casserole. Sprinkle with marinade and refrigerate 1 hour or more. Separate onions into rings and place 2 or 3 on each pattie. Place 2 strips pepper on each pattie, forming an "X". Cover with wax paper and **MW on 70% (Medium-high) 5 to 6 minutes,** turning each pattie halfway around after 3 minutes. Place a slice of pineapple on each pattie and let stand 3 minutes. Serve with sauce on hamburger buns, if desired. Makes 4 servings.

BURGER LOGS

1 pound ground beef
¼ cup barbecue sauce
¼ cup fine bread crumbs
Kitchen Bouquet
4 hot dog buns
Shredded cheddar cheese

Mix ground beef, barbecue sauce and bread crumbs. Divide mixture into four equal portions; roll each portion into a log shape. Brush all sides lightly with Kitchen Bouquet; place on a meat or bacon rack and cover with wax paper. **MW on 70% (Medium-high) 6 to 7 minutes,** turning logs over midway through cooking. *(A pair of tongs make this easy.)* Place logs in hot dog buns and sprinkle cheese over top to serve. Makes 4 servings.

▶ *Hamburgers*

PIZZA JOE

- 1 pound ground beef
- 1 (1½-ounce) packet spaghetti sauce mix
- 1 teaspoon Italian seasoning
- 1 (6-ounce) can tomato paste
- 1 cup water
- 8 hamburger buns

Crumble ground beef into a hard-plastic colander set in a 1½-quart round casserole. **MW on High 5 minutes,** stirring midway through cooking. Discard grease and transfer meat into same casserole. Stir in spaghetti sauce mix, Italian seasoning, tomato paste and water. Cover and **MW on High 6 minutes,** stirring midway through cooking. Spoon into buns to serve. Makes 8 servings.

BALONEY BURGERS

- 8 ounces bologna
- 1 cup shredded cheddar cheese (4 ounces)
- 2 tablespoons prepared mustard
- 2 tablespoons pickle relish
- 2 tablespoons mayonnaise
- Margarine
- 6 hamburger buns

Process bologna in food processor till fine. Add cheese, mustard, pickle relish and mayonnaise. Blend. Spread margarine on cut sides of buns. Divide filling mixture among buns. Wrap each bun in paper toweling. For each bun, **MW on 70% (Medium-high) 25 to 30 seconds.** Makes 6 servings.

"Please send me a copy of your book—I'm tired of having to borrow one!"
Mrs. Jack M. Covin, Longview, Texas

HOT DOGS

Hot Dog! Who hasn't microwaved one? Since frankfurters are fully cooked, they take little time to heat and are perfect for microwaving.

Frankfurters are a type of sausage which originated in Frankfurt, Germany. "Hot dogs" and "Wieners" are names used interchangeably for frankfurters. The existence of sausage has been traced to 900 B.C., and different types can be found in every country of the world.

Bratwurst, knockwurst and Polish sausage are also fully cooked and great to microwave. Since they are larger, they take longer to heat than the average frankfurter.

HOW TO DEFROST FRANKFURTERS

Frankfurters can be frozen in their original packaging for several months. Because they are fully cooked, they defrost rapidly.

If you plan **to heat them immediately,** defrost them on High power. Put the package on a paper towel or plate in the microwave oven. **MW on High 1 minute.** Remove the packaging and separate the hot dogs, if possible.

MW on High 1 minute. Rearrange the hot dogs so that the most frozen ones are on the outside. **MW on High 1 minute** more. Heat as desired.

If you don't need the frankfurters for several hours, transfer the frozen package from the freezer to the refrigerator. This saves the energy used in microwave defrosting.

HOW TO DEFROST HOT DOG BUNS

Remove the metal twist-tie from the plastic bag and slip a double layer of paper toweling between the buns and the plastic bag, both top and bottom. Place bag on a microwave meat or bacon rack. For a package of **8 to 10 hot dog buns, MW on 30% (Medium-low) 2½ to 3 minutes.** Rotate package top to bottom midway through defrosting. Redistribute each layer of buns so that the least defrosted areas are moved to the outside. Separate buns and let stand to finish defrosting.

MICROWAVING HOT DOGS IN BUNS

You've probably already discovered that if you microwave hot dogs in buns too long, the buns will become tough, chewy and/or soggy. Besides reducing cooking time, there are several ways to prevent this and microwave franks to perfection.

- Do not spread buns with mustard, ketchup or relish before heating. Condiments will make buns soggy.
- **Pierce hot dogs** with a fork or knife to prevent steam buildup and exploding.
- Place the hot dogs in buns and **wrap individually in paper towels** or napkins to absorb steam.
- Arrange wrapped hot dogs on a microwave meat or bacon rack so that air can circulate around them.
- Do not microwave more than four wrapped hot dogs at a time, because they will heat unevenly.
- Microwave according to time chart below for hot dogs 10-per-pound size:

Number of Hot Dogs in Buns	Minutes on High Power
1	½ to ¾
2	¾ to 1¼
3	1¼ to 1¾
4	1¾ to 2¾

- Hot dogs larger or smaller than 10 per pound will vary in heating time. Increase or subtract time accordingly.
- Hot dogs manufactured with cheese inside take the same time to microwave as meat wieners.
- For jumbo franks (4 per pound), bratwurst, knockwurst, and Polish sausages, we recommend heating them in a dish separately before placing in buns.
- Sausages in a thick casing must be pierced several times and heated on paper toweling to absorb the grease.
- We recommend heating canned chili hot dog sauce separately in a dish. Stir once or twice during heating, and cover to avoid splattering due to fat content.

Who says hot dogs taste better at the ballpark? Try some of our variations if hot dogs have gotten boring at your house!

OSCAR J. REUBENS

1 pound smoked frankfurters (8 per pound)
8 hot dog buns
1 (8-ounce) can sauerkraut, well drained
Caraway seeds
Thousand Island dressing
Shredded or sliced Swiss cheese

Pierce each frankfurter and place into split buns. Arrange 2 tablespoons sauerkraut over each frank. Sprinkle with caraway seeds. Drizzle with dressing and add desired amount of cheese. Arrange four Reubens on a paper towel covered rack. **MW on High 2 to 3 minutes,** or until heated through. Repeat with remaining four. Makes 4 to 8 servings.

CONEY ISLAND HOT DOGS

½ pound ground beef
1 medium onion, chopped
1 (8-ounce) can tomato sauce
1 teaspoon chili powder
½ teaspoon worcestershire sauce
1 (16-ounce) package frankfurters, 10 in package
10 hot dog buns

Break up ground beef into a 4-cup glass measure. Sprinkle onions over top. **MW on High 3 to 4 minutes,** breaking up meat midway through cooking. Pour off grease. Add tomato sauce, chili powder and worcestershire sauce. **MW on High 4 minutes.** Pierce franks and place in buns. Heat according to chart on page 77. Top with sauce and serve. Makes 5 to 10 servings.

JEFF'S BACON CHEESE DOGGIES

The youngest Steiner's favorite

Bacon slices
Frankfurters
Sharp cheddar cheese
Hot dog buns

Use one bacon slice for each frankfurter. Place bacon on paper toweling either on a bacon rack or paper plate. Cover with a paper towel. **MW on High about 45 seconds per slice,** or until almost crisp. Cut a lengthwise pocket in frankfurter and stuff with a strip of cheese. Wrap bacon around doggie and secure with toothpicks. Place in split hot dog buns and microwave according to chart on page 77. *Serve with condiments of your choice.*

Quick-Tip ➡ If the interior of your microwave oven gets stained, clean with non-abrasive liquid cleanser (such as Soft Scrub).

CHIHUAHUA DOGS

1 (10-ounce) can chili hot dog sauce
1 (16-ounce) package frankfurters, 10 in package
10 pre-formed taco shells
  Chopped onion
  Shredded cheese
  Chopped tomato and lettuce (optional)

To heat sauce, transfer to a microwave-safe dish and **MW on High 2 to 3 minutes,** stirring once midway through cooking. Pierce franks and place one in each taco shell. Top with heated chili sauce, onion and cheese as desired. Place on paper towels and microwave according to chart on page 77. Sprinkle with chopped tomato and lettuce, if desired. Makes 5 to 10 servings.

HOPSCOTCH HOT DOG CASSEROLE

1 (2-pound) bag frozen Southern-style hash-brown potatoes
½ cup chopped onion
2 tablespoons margarine
2 tablespoons flour
¼ teaspoon garlic powder
  Salt and pepper to taste
1¼ cups milk
1 cup shredded cheddar cheese (4 ounces)
6 frankfurters, sliced ¼-inch thick
  Paprika

Make a 1-inch slit in bag of potatoes and place bag on a paper plate. **MW on High 13 to 15 minutes,** turning bag upside down and redistributing contents every 5 minutes. Let stand 5 minutes.

Place onion and margarine in a 4-cup glass measure. **MW on High 2 to 2½ minutes.** Stir in flour, garlic powder, salt and pepper. Blend in milk with a whisk. **MW on High 3 to 4 minutes,** whisking midway through cooking. Stir in cheese until melted.

Combine potatoes and frankfurters in a 2-quart round casserole. Pour cheese sauce over top and blend. Dust top with paprika. Cover and **MW on 70% (Medium-high) 6 to 8 minutes,** rotating dish midway through cooking. Let stand 5 minutes before serving. Makes 6 to 8 servings.

MICROWAVE TO GRILL

Have you ever charcoaled chicken or spareribs which ended up tasting like charcoal themselves? Eliminate this problem by precooking them in the microwave oven.

By combining your outdoor grill with your microwave oven, meats will be juicer with less chance of overcooking. You can also precook vegetables that are not on the grill long enough to become tender.

Steaks, meat for shish kabobs, and hamburger patties do not need precooking, and are perfect for outdoor grilling. You can cook extras to freeze "for a rainy day." Defrost and reheat them in the microwave oven for a "just-grilled" flavor, see page 81.

For chicken (whole, halves or pieces), Cornish hens, spareribs, roasts and turkey (whole, breast or hindquarter), precook in the microwave oven on **High power 5 minutes per pound.** Turn top to bottom midway through cooking. Pour off accumulated juices before grilling so you won't douse the flames! Transfer the precooked foods to the grill while they're hot for continuous cooking, see page 81.

While the food is microwaving, light your charcoal or gas grill. Usually it takes about 30 minutes for charcoal to be ready, and about 10 minutes to preheat a gas grill.

To test the degree of heat from your fire, count the number of seconds you can comfortably hold your hand over the fire. One or two seconds is a hot fire (good for steaks, burgers and kabobs); three or four seconds is a medium fire (good for chicken, Cornish hens and spareribs); five or six seconds is a slow fire (good for turkeys and roasts).

Since foods will be nearly done when precooking in the microwave oven, generally they will be done when the desired smoking and browning has occurred. Cover the grill for optimum smoke flavor.

To **precook baking potatoes for grilling,** pierce skin once, arrange on a meat rack, and **MW on High 4 to 5 minutes per pound.** Wrap in foil, or leave natural, and place on grill for 15 to 20 minutes, turning top to bottom midway through grilling.

Fresh **corn** is wonderful for grilling. Remove husks and silk. Place in a microwave-safe dish and cover with plastic wrap. **MW on High 2 minutes per ear.** Spread with butter and wrap each ear individually in aluminum foil. Place on hot grill for 10 minutes.

You can be energy efficient by grilling extra food while the flames are still hot. For hamburgers, undercook slightly, remove from grill and let cool. To package for the freezer, stack hamburgers between squares of wax paper and wrap in suitable freezer material. When you need a meal in a hurry, separate the number of **frozen patties** you need by wedging a dinner knife between the squares of wax paper. **MW on 70% (Medium-high) until defrosted and heated** as desired. They will taste as charcoaled as the day you grilled them!

For grilling extra steaks, sear them over a hot fire and leave them rare. Allow them to cool and wrap for the freezer. To prepare for serving, remove wrapping and place **frozen steak(s)** on a microwave meat or bacon rack. **MW on 50% (Medium) until defrosted and heated as desired.**

MICROWAVE GRILLED SPARERIBS

3 to 3½ pounds country style spareribs
1 cup water
Bottled barbecue sauce or your own favorite recipe

Arrange spareribs in a single layer in a 2-quart rectangular dish. Pour water over spareribs and cover with wax paper. **MW on High 15 minutes,** turning ribs over midway through cooking. Pour off water and accumulated juices.

Grill ribs over a medium fire for 30 minutes, basting often with barbecue sauce and turning as needed. Keep grill covered for maximum smoke flavor. Makes 4 servings. Reheat frozen or refrigerated ribs on 50% (Medium).

MICROWAVE GRILLED CHICKEN

Chicken pieces (number of pounds as desired)
Bottled barbecue sauce or your own favorite recipe

Arrange chicken pieces in a single layer in a 2 or 3-quart rectangular dish. Place thick pieces near the outside of the dish, and thin or bony pieces in the center. Cover with wax paper and **MW on High 5 minutes per pound,** turning pieces top to bottom midway through cooking. Pour off accumulated juices.

Grill chicken pieces over a medium fire for 20 minutes, basting often with barbecue sauce and turning as needed. Keep grill covered for maximum smoke flavor.

Freeze or refrigerate extra pieces for future meals. To reheat one piece of home-frozen chicken, **MW on 70% (Medium-high) 2½ to 3 minutes.**

A refrigerator-temperature chicken piece will reheat in 1½ to 2 minutes. Turn each piece top to bottom midway through reheating. Time needed for reheating depends on the size and number of pieces.

Sauces for Meat

MUSHROOM BORDELAISE SAUCE

¼ cup flour
¼ cup oil
¾ cup chopped onion
1 (10¾-ounce) can condensed beef bouillon
⅓ cup Bordeaux or red wine
8 ounces fresh mushrooms, cleaned and sliced
Pepper to taste

To make brown roux: Blend flour and oil in a 4-cup glass measure. **MW on High 4 minutes;** stir. Stirring after each minute, **MW on High 1 minute at a time,** until brown. When desired color is reached, immediately add onions to halt further browning. **MW on High 3 minutes** to cook onions.

Stir in bouillon and **MW on High 2 to 3 minutes,** or until mixture comes to a boil and is thickened. Add wine.

Place mushrooms in a small casserole. Cover with plastic wrap and **MW on High 3 to 4 minutes,** or until cooked. Drain and add to sauce mixture. Reheat if necessary. Yields 2 cups.

CINDY'S MUSHROOM SAUCE
An inspiration of one of Ann's two daughters

8 ounces fresh mushrooms, cleaned and sliced
2 tablespoons margarine
1 tablespoon flour
1 teaspoon dry mustard
¼ teaspoon marjoram
⅛ teaspoon pepper
1 teaspoon instant beef bouillon granules
1 cup milk
½ cup shredded Swiss cheese

Combine mushrooms and margarine in a 4-cup glass measure. Cover and **MW on High 3 to 4 minutes,** stirring midway through cooking. Remove mushrooms with slotted spoon and place on paper plate. Set aside.

Add flour, mustard, marjoram, pepper and bouillon granules to liquid from mushrooms. Use whisk to blend in milk. Whisking midway through cooking, **MW on High 2½ to 3 minutes,** or until thickened.

Stir in cheese and reserved mushrooms. **MW on 70% (Medium-high) 1½ to 2 minutes,** or until heated through. Serve over hamburgers or steak. Yields 1½ cups.

MICROWAVING MEAT LOAVES

Can you bake a meat loaf in 12 minutes or less? You surely can if you microwave it! Even a two-pound meat loaf takes only 20 minutes to cook in the microwave oven.

Here are some tips for microwaving a tasty meat loaf:

- If using the microwave oven to defrost frozen ground meat to form a meat loaf, flake off thawed meat periodically and remove from oven so that it won't begin to cook. The meat should still be cool to touch. Meat that is warm will facilitate bacteria action to take place much quicker.
- Beat eggs well before adding to meat loaf mixture. Eggs that aren't well beaten tend to leave coagulated white areas in the microwaved meat loaf, which doesn't enhance its appearance.
- Since meat loaves microwave so quickly, ingredients such as chopped onions and green pepper tend to remain quite crisp. If you desire a softer texture to these ingredients, place the chopped onion/pepper in a glass measure and cover with plastic wrap. **MW on High 2 to 3 minutes per cup** of vegetables before adding to other meat loaf ingredients.
- Unless a recipe specifies lean ground beef, regular ground beef is a better buy for making a meat loaf according to Merle Ellis in his syndicated newspaper column, "The Butcher."
- Ground meats, such as pork, lamb and ham, can be used to make a meat loaf. However, combining these with some ground beef gives a better texture.
- Place the meat loaf on a meat or bacon rack, which has been sprayed with Pam. The fat in the meat loaf will drain away from the meat loaf as it microwaves.
- Large meat loaves, those made with more than 1½ pounds of meat, will microwave more uniformly if shaped into a wreath. The open area in the center will allow microwave energy to penetrate from the center as well as the outer edges.
- In shaping a loaf, it is helpful to have the loaf flat rather than domed in the center. If the ends of the meat loaf show signs of over-cooking, shield them with pieces of aluminum foil (see page 183).
- If using a meat probe, insert the probe horizontally through one end of the meat loaf to the center. Microwave the meat loaf to 160°.
- If desired, additional browning can be created by brushing the top of the meat loaf with a browning sauce, such as Kitchen Bouquet, or sprinkling Micro Shake on the top and sides.
- Cover the meat loaf with wax paper to hold in heat for more even cooking.
- Use 70% (Medium-high) rather than High power to eliminate cracks and lessen the shrinkage.
- Standing time is important after a meat loaf has finished microwaving. Let a meat loaf stand at least 5 minutes and as long as one-half the time it took to microwave. It'll slice easier after standing.

Meat Loaf

ANN LANDERS' MEAT LOAF

MicroScope received permission to convert Ann Landers' Meat Loaf recipe. Our adaptation is her favorite meat loaf recipe from one of her columns.

- 2 **pounds ground round steak**
- 2 **eggs, well beaten**
- 1½ **cups dry bread crumbs**
- ¾ **cup catsup**
- ½ **cup warm water**
- 1 **(4-serving size) envelope dry onion soup mix**
- 1 **(8-ounce) can tomato sauce**

Combine all ingredients except tomato sauce and mix thoroughly. Place mixture in a 10-inch glass pie plate and shape into a wreath, leaving a 2-inch open area in the center. Cover with wax paper and **MW on 70% (Medium-high) 16 to 17 minutes,** rotating dish twice during cooking. Use a baster to remove any grease that may have collected in the bottom of the plate.

Pour tomato sauce over top and **MW (uncovered) on 70% 4 to 6 minutes,** rotating dish once. Let stand 5 to 7 minutes before slicing. Makes 6 to 8 servings.

Quick-Tip ➡ Ressurrect ballpoint pens that no longer write. Use this procedure for a one-piece plastic ballpoint pen or a plastic ink tube which can be removed from a refillable pen. The small amount of metal in the writing tip causes no problems in the microwave oven. Wrap plastic pen or ink tube in a piece of plastic wrap which will prevent ink from squirting the oven. Place ½-cup water in a 1-cup glass measure. Lay wrapped pen or tube across top. **MW on High 20 to 30 seconds.** Scribble on scratch paper to activate ink flow.

→ Meat Loaf

CHEESY MEAT LOAF

- ½ cup chopped onion
- ¼ cup chopped green bell pepper
- 1½ pounds ground beef
- 1 cup Pepperidge Farm stuffing mix
- 1 (8-ounce) can seasoned tomato sauce
- 2 eggs, well beaten
- ¼ teaspoon thyme
- ¼ teaspoon marjoram
- ½ cup shredded Swiss cheese

Combine onion and green pepper in a 1-cup glass measure. Cover and **MW on High 1½ to 2 minutes.** Add onion and green pepper to remaining ingredients; mix well. Shape meat mixture into loaf and place in a 9x5x3-inch glass loaf pan. Cover with wax paper and **MW on 70% (Medium-high) 15 to 17 minutes,** rotating pan twice during cooking. Let stand 5 to 7 minutes before slicing. Makes 6 to 8 servings.

14-KARAT MEAT LOAF

- ½ cup chopped onion
- ⅔ cup shredded carrots
- ⅔ cup quick oats, uncooked
- ¼ cup ketchup
- 1 egg, well beaten
- 1 teaspoon worcestershire sauce
- ¼ teaspoon pepper
- 1 pound ground beef

Place onion in a 1-cup glass measure. Cover and **MW on High 1½ to 2 minutes.** In a medium bowl, combine onions with remaining ingredients. Shape meat mixture into a uniform loaf and place on a microwave meat or bacon rack which has been sprayed with Pam. Cover with wax paper and **MW on 70% (Medium-high) 9 to 10 minutes,** rotating rack midway through cooking. Let stand 5 minutes before slicing. Makes 4 to 5 servings.

NOTE: *To double recipe, shape meat mixture into a wreath, leaving a 2-inch open area in the center.* **MW on 70% 18 to 20 minutes.**

MINI-MEX MEAT LOAVES

- 4 (6-inch) corn tortillas
- 1 pound lean ground beef
- ½ cup dry bread crumbs
- ⅓ cup picante sauce
- 1 egg, well beaten
- 1 teaspoon chili powder
- Shredded cheddar cheese, chopped lettuce and tomatoes

In outer edge of each tortilla, make 4 (2-inch) cuts toward center. (Make a cut at 3, 6, 9 and 12 o'clock positions.) Place a tortilla in each of 4 (10-ounce) custard cups.

Combine meat, bread crumbs, picante sauce, egg and chili powder; blend well. Divide meat mixture into four equal portions. Press each meat portion into a tortilla-lined custard cup. Make an indentation in center of each mini loaf. Place cups on a microwave-safe tray; cover with wax paper. **MW on 70% (Medium-high) 8 to 9 minutes,** rotating cups midway through cooking. Top each meat loaf with cheese and let stand 2 to 3 minutes. Garnish with lettuce and tomatoes to serve. Makes 4 servings.

HERBED MINI-LOAVES

- 1 pound ground beef
- 1 cup soft bread crumbs
- 1 egg, well beaten
- 1 clove garlic, minced
- ½ onion, chopped
- ¼ cup snipped fresh parsley
- 1 teaspoon salt
- ½ teaspoon dried oregano
- ½ teaspoon dried basil
- ¼ teaspoon pepper
- 1 (8-ounce) can tomato sauce

Combine ground beef, bread crumbs, egg, garlic, onion, parsley, salt, oregano, basil, pepper and half of sauce, blend well. Shape meat mixture into 6 balls and place each into individual space in a microwave muffin pan. *(If using microwave muffin pan with steam vent holes, set pan on several layers of paper toweling on a tray or dinner plate to absorb grease as it drains.)* **MW on 70% (Medium-high) 11 to 13 minutes,** rotating pan midway through cooking. Drizzle with remaining sauce and let stand 5 minutes before serving. Makes 4 to 6 servings.

> **Meat Loaf**

LAMB LOAF

1½ pounds ground lamb
2 pieces of bread, crumbed
2 green onions with tops, thinly sliced
1 egg, well beaten
2 tablespoons water
⅛ teaspoon allspice

Combine all ingredients; mix well. Shape in loaf and place on a microwave meat or bacon rack which has been sprayed with Pam. Cover with wax paper. **MW on 70% (Medium-high) 11 to 12 minutes,** rotating rack midway through cooking. Let stand 5 to 7 minutes before slicing. Makes 5 to 6 servings.

TURKEY MEAT LOAF

1 medium green bell pepper
½ cup chopped onion
½ pound ground cooked turkey
1 pound ground beef
¾ cup seasoned bread crumbs
2 eggs, well beaten
Ketchup

Slice 3 rings from bell pepper. Chop remaining pepper and combine with onion in a 2-cup glass measure. Cover and **MW on High 3 to 3½ minutes,** stirring midway through cooking. Add to ground turkey and beef, bread crumbs, eggs and ½ cup ketchup. Shape meat mixture into loaf and place in a 9x5x3-inch glass loaf pan. Cover with wax paper and **MW on 70% (Medium-high) 11 to 12 minutes,** rotating pan midway through cooking. Drizzle loaf with ketchup and top with bell pepper rings. **MW (uncovered) on High 1½ minutes.** Let stand 5 to 7 minutes before slicing. Makes 6 servings.

"Dear CiCi and Ann,
Thank you so much for writing your book. I like it so much that I would like my mother, a new microwave oven owner, to have it."

More 'power' to you
Vickie Sherman, St. Louis, Missouri

CHILI NOODLE DINNER

- 1 pound ground beef
- 4 cups uncooked medium egg noodles
- 1 (1¾-ounce) packet chili seasoning mix
- 1 (16-ounce) can sliced stewed tomatoes
- 1½ cups water OR 1 (12-ounce) can light beer

Crumble ground beef into a hard-plastic colander set in a 3-quart round casserole. **MW on High 5 minutes,** stirring midway through cooking. Set meat aside. Discard grease.

Place noodles in bottom of same casserole. Distribute meat over top of noodles. Sprinkle seasoning mix over meat. Pour tomatoes and water over top. Cover and **MW on High 12 to 14 minutes,** stirring midway through cooking. Let stand 5 to 7 minutes before serving. Makes 6 servings.

STUFFED SPAGHETTI SQUASH

- 1 (3-pound) spaghetti squash
- 1 pound ground beef
- 1 (4-ounce) can mushroom pieces, drained
- 2 (8-ounce) cans tomato sauce
- 1 teaspoon salt
- ¼ teaspoon pepper
- ¼ teaspoon basil
- ¼ teaspoon oregano
- ⅛ teaspoon garlic powder
- ¼ cup grated parmesan cheese

To cook squash: Using an icepick, make 8 holes in squash at intervals around it. Weigh squash and determine microwave time at 6 to 7 minutes per pound. Place squash in a 2-quart rectangular dish. **MW on High,** turning squash upside down midway through cooking. Let stand.

To make sauce: Crumble ground beef into a hard-plastic colander set in a 3-quart round casserole. **MW on High 5 minutes,** stirring midway through cooking. Discard grease and transfer meat into same casserole. Add mushrooms, tomato sauce, salt, pepper, basil, oregano and garlic powder.

To stuff squash: Slice squash in half lengthwise. Remove seeds and scoop strands of squash into casserole with sauce; combine thoroughly. Place the two squash shells back into the rectangular dish. Divide meat mixture and spoon into the shells. Sprinkle parmesan cheese over mixture. Cover with wax paper and **MW on High 8 to 10 minutes,** or until mixture is heated through. Makes 6 servings.

Ground Beef

TEX-MEX TORTILLA CASSEROLE

- 1 pound ground beef
- 1 large onion, chopped (2 cups)
- 1 (16-ounce) can cream-style corn
- 1 (16-ounce) can stewed tomatoes
- 1 (10-ounce) can enchilada sauce
- 12 corn tortillas
- 2 cups shredded sharp cheddar cheese (8 ounces)

Crumble ground beef into a hard-plastic colander set in a 3-quart round casserole. Sprinkle onion on top. **MW on High 6 to 7 minutes,** stirring midway through cooking. Discard grease and transfer mixture into same casserole. Add corn, tomatoes (including liquid) and enchilada sauce. Stir well.

Tear four tortillas into 1-inch pieces and place on bottom of a 2-quart rectangular dish. Pour ⅓ of meat mixture evenly over tortillas, and sprinkle with ⅓ of the cheese. Repeat for two more layers, reserving cheese for top of casserole. Cover and **MW on High 10 minutes,** rotating dish midway through cooking. Uncover and sprinkle reserved cheese over top of casserole. **MW on High 2 minutes,** or until cheese melts and casserole is bubbly hot. Makes 8 servings.

NOTE: *Freezes well. This recipe can be divided into two (1-quart) casseroles, serving 4 each. For freezing directions, see page 101.*

GAUCHO CASSEROLE

- 1 pound ground beef
- 1 (1¼-ounce) packet taco seasoning mix
- 1 cup water
- 1 (15-ounce) can chili beans in sauce
- 1 (16-ounce) can sliced stewed tomatoes
- 1⅓ cups instant rice

Crumble ground beef into a hard-plastic colander set in a 3-quart round casserole. **MW on High 5 minutes,** stirring midway through cooking. Discard grease and transfer meat into same casserole. Add seasoning mix, water, beans in sauce, and stewed tomatoes. Cover and **MW on High 6 to 7 minutes,** or until bubbly hot.

Stir rice into hot mixture. Cover and let stand 10 minutes for rice to absorb liquid. Makes 6 servings.

Ground Beef

CHILI FROM A MIX

- 2 pounds ground beef
- 1 (3-ounce) packet chili mix
- 1 (8-ounce) can tomato sauce
- 12 ounces beer or water

Crumble ground beef into a hard-plastic colander set in a 3 to 4-quart round casserole. **MW on High 8 to 9 minutes,** stirring midway through cooking. Discard grease and place meat into same casserole. Add chili mix and stir until well combined. Add tomato sauce and beer or water. Cover and **MW on High 20 minutes,** stirring midway through cooking. Let stand 10 minutes before serving. Leftover chili can be reheated on High power. Makes 6 to 8 servings.

Serving Suggestion: *For hotter taste, add ground red chilies, Tabasco sauce, or jalapeños.*

TACO CASSEROLE SAN ANTONE

In honor of our SAN ANTONIO LIGHT

- 1 pound ground beef
- 1 medium onion, chopped
- 1 cup milk
- 2 eggs
- 2 teaspoons chili powder
- ½ teaspoon ground cumin
- 1 (8-ounce) can tomato sauce
- 3 cups regular-size corn chips
- 1 (10-ounce) package Monterey Jack cheese with jalapenos, shredded
- 1 cup sour cream

Crumble ground beef into a hard-plastic colander set in a 3-quart round casserole. Sprinkle onion on top. **MW on High 6 to 7 minutes,** stirring midway through cooking. Discard grease and set meat aside.

In a 4-cup glass measure, combine milk, eggs, chili powder and cumin. Beat well. Slowly add tomato sauce, beating continuously. Place half of corn chips in same 3-quart casserole. Top with half of meat mixture, 1 cup of cheese, and half of tomato sauce mixture. Repeat layers, reserving ½ cup of cheese.

MW on 70% (Medium-high) 10 to 12 minutes, rotating dish midway through cooking. Spread sour cream over casserole and sprinkle with reserved cheese. **MW on 70% 2 to 4 minutes,** or until cheese is melted. Makes 6 servings.

Ground Beef

TORTILLA LASAGNE

- 1½ pounds ground beef
- 1 cup chopped onion
- ½ teaspoon pepper
- 1 teaspoon salt
- ½ teaspoon garlic powder
- 1½ teaspoons cumin
- 4 teaspoons chili powder
- ¾ cup water
- 1 cup taco sauce
- 12 corn tortillas
- 1 cup sour cream
- 3 cups shredded Monterey Jack cheese (12 ounces)

Crumble ground beef into a hard-plastic colander set in a 2-quart casserole. Sprinkle onion on top. **MW on High 8 minutes,** stirring midway through cooking. Discard grease and transfer meat mixture into same casserole. Stir in pepper, salt, garlic powder, cumin, chili powder and water. Cover and **MW on High 5 minutes.** Pour ¼ cup taco sauce in bottom of a 2-quart rectangular casserole. Place 6 tortillas in dish. Layer the remaining ingredients in this order:

- ¼ cup taco sauce
- meat mixture
- sour cream
- half of cheese
- ¼ cup taco sauce
- 6 tortillas
- ¼ cup taco sauce
- remaining half of cheese

Dust with additional chili powder for color. Cover with plastic wrap and **MW on High 10 minutes,** or until heated through. Makes 8 servings.

Quick-Tip ➡ Instead of being a member of the "clean the plate and bowl club," package leftover entrees in either plastic bags designed for freezing or in a microwave-safe divided utensil to reheat for another meal.

POCKET TACOS

1 pound ground beef
1 (1¼-ounce) packet taco seasoning mix
½ (15-ounce) can refried beans
½ cup water
1 package of 6 pita pocket bread loaves
1 cup shredded cheddar cheese (4 ounces)
Shredded lettuce
Chopped tomatoes

Crumble ground beef into a hard-plastic colander set in a 1½-quart round casserole. **MW on High 5 minutes,** stirring midway through cooking. Discard grease and transfer meat into same casserole. Stir taco mix, beans and water into meat. **MW on High 6 minutes,** stirring midway through cooking.

Cut each pita bread in half "across the equator." Carefully open pockets and put some of the meat mixture and cheese into each one. To warm six pockets just before serving, **MW on High 30 to 40 seconds.** Add lettuce and tomatoes. Makes 12 pockets.

VEGETABLE MEAT PIE

1 (10-ounce) package frozen mixed vegetables
1 pound very lean ground beef
1 cup Wheat Chex, crushed to yield ½ cup (no substitution)
1 egg
⅓ cup ketchup
¼ cup finely-chopped onions
1½ teaspoons Worcestershire sauce
1 teaspoon salt
½ teaspoon chili powder
½ cup milk
2 large or extra-large eggs
1 cup shredded sharp cheddar cheese (4 ounces)
¼ teaspoon dill weed

Place package of frozen vegetables on a paper towel or plate on the floor of the microwave oven. **MW on High 5 minutes.** Set aside.

To make meat crust, combine ground beef, Wheat Chex, egg, ketchup, onions, worcestershire sauce, salt and chili powder. Press evenly on bottom and sides of a 10-inch glass pie plate.

Beat together milk and eggs. Stir in reserved vegetables, cheese and dill weed. Pour into uncooked meat crust. Rotating several times during cooking, **MW on 70% (Medium-high) 18 to 22 minutes,** or until center barely jiggles. Let stand 10 minutes before cutting. Makes 5 to 6 servings.

Ground Beef

BACON AND BEEF SUCCOTASH

- 2 tablespoons bacon drippings
- 1 pound lean ground beef
- ¾ cup chopped onion
- 2 (16-ounce) cans succotash
- 1 (10¾-ounce) can cream of celery soup
- 6 slices bacon, cooked and crumbled
- Pepper to taste
- 1 cup crushed potato chips or corn flakes
- Paprika

Combine bacon drippings, ground beef and onion in a 2-quart round casserole. **MW on High 6 minutes,** stirring midway through cooking.

Drain and discard liquid from 1 can of succotash. Add contents of drained can plus liquid and contents of second can of succotash to meat mixture. Blend in soup, bacon and pepper. **MW on High 4 to 5 minutes.** Stir and top with crushed chips. Sprinkle paprika over top. **MW on High 3 to 4 minutes,** or until heated through. Makes 5 servings.

HAMBURGER STEW

Stuff it in a baked potato to make a meal.

- 1 pound ground beef
- ½ cup chopped onion
- 1 (1½-ounce) package Beef Stew seasoning mix
- 2 tablespoons flour
- 1¼ cups water
- ¼ cup dry red wine
- 1 cup frozen peas and carrots
- 4 to 5 large potatoes, baked

Crumble ground beef into a hard-plastic colander set in a 2-quart round casserole. Sprinkle onion on top. **MW on High 6 to 7 minutes,** stirring midway through cooking. Discard fat and set meat mixture aside.

Place seasoning mix and flour in 2-quart casserole. Blend in water and wine with a wire whisk. Add meat. **MW on High 5 minutes,** stirring once. Add peas and carrots and **MW on High 3½ to 4½ minutes,** or until vegetables are desired doneness. Serve as a potato topper. Makes 4 to 5 servings.

Beef

BURGUNDY STEAK STRIPS

- 1 (¾-ounce) packet beef mushroom gravy mix
- ½ cup burgundy wine
- ½ cup water
- 1 pound flank steak
- 1 tablespoon worcestershire sauce
- 2 (16-ounce) cans small whole potatoes, drained and quartered
- 1 tablespoon snipped parsley

Combine gravy mix with wine and water in a 2-cup glass measure. **MW on High 2 to 3 minutes,** stirring with whisk midway through cooking. After gravy has thickened, set aside.

Slice flank steak diagonally across grain into thin slices. Place in a 2-quart casserole. Cover with wax paper and **MW on High 4 minutes,** stirring midway through cooking. Stir in reserved gravy, worcestershire sauce and potatoes. Re-cover with wax paper and **MW on High 4 to 6 minutes,** or until heated through. Stir and garnish with parsley. Makes 4 to 5 servings.

COWPUNCHERS' BEEF STEW

In honor of our FORT WORTH STAR-TELEGRAM

- 1½ pounds potatoes, pared and cut into small pieces
- 5 carrots, cut into coins
- 1 rib celery, sliced thin diagonally
- Water
- 1 (16-ounce) jar boiled whole onions, drained
- 2 to 3 cups slivered cooked roast

Place potatoes, carrots and celery in a 2-quart glass batter bowl. Add 1 cup of water to vegetables and cover with plastic wrap. **MW on High 16 to 18 minutes,** or until vegetables are tender.

Remove vegetables with a slotted spoon and set aside. To remaining liquid from vegetables, add leftover meat stock from roast. Make 2 cups of Homemade Gravy, page 95.

Add reserved vegetables, onions and roast to gravy. **MW on High until reheated.** Add salt and pepper as desired. Makes 6 servings.

Beef

HOMEMADE GRAVY

Determine amount of flour for thickening gravy by using 1 tablespoon flour per cup of broth. In a small bowl, combine flour with an equal amount of water and stir until smooth. Using a whisk, blend flour mixture into liquid. Add Kitchen Bouquet to make gravy desired brown color. **MW on High until mixture begins to boil,** whisking midway through cooking.

ROUND STEAK ROLL-UPS

- 1 (4-ounce) can mushroom pieces, drained and chopped
- ¼ cup chopped onion
- ⅔ cup shredded Jarlsburg cheese
- ⅓ cup grated parmesan cheese
- ⅔ cup fine bread crumbs
- 1 tablespoon parsley flakes
- Salt and pepper to taste
- 1½ pounds round steak
- 2 tablespoons dry red wine
- 1 tablespoon Kitchen Bouquet
- Water
- 1 (⅞-ounce) packet brown gravy mix

Reserve ¼ cup mushrooms and combine remaining with onions in a 4-cup glass measure. **MW on High 2 minutes.** Add cheeses, bread crumbs, parsley, salt and pepper; mix thoroughly. Set aside.

Pound steak to ¼-inch thickness and divide into 6 pieces. Combine wine and Kitchen Bouquet and drizzle over one side of steak pieces. Turn steak over and divide filling mixture among pieces. Roll each piece pinwheel fashion and secure end with toothpicks. Place rolled steaks seam side up in a wagon-wheel arrangement on a microwave-safe platter or tray. Cover with wax paper and **MW on 50% (Medium) 15 to 17 minutes,** turning steaks over and rotating dish midway through cooking. Let stand 8 minutes.

Pour juices that have collected into a 2-cup glass measure. Add water to make 1 cup liquid. Stir gravy mix into liquid and **MW on High 2 to 2½ minutes,** or until mixture bubbles. Add reserved mushrooms and **MW on High 1 to 2 minutes,** or until gravy is thickened. Serve with Round Steak Roll-Ups. Makes 6 servings.

BEEF BOURGUIGNONNE

   6 slices bacon, quartered
   2 pounds boneless sirloin, cut into ¾-inch cubes
   ¼ cup flour
   8 ounces fresh mushrooms, sliced
   1 medium onion, cut in eighths
   1 clove garlic, minced
   1 bay leaf
   1 tablespoon fresh snipped parsley
   ½ teaspoon thyme
1¼ cups burgundy wine
   2 teaspoons instant beef bouillon granules
      Salt and pepper to taste

Place bacon in a 4-quart simmer pot. Cover and **MW on High 4 to 5 minutes.** Toss sirloin cubes with flour and add to bacon and drippings. Toss to coat with drippings. Sprinkle any remaining flour over meat. Add mushrooms, onion, garlic, bay leaf, parsley and thyme. Add wine, bouillon, salt and pepper. Cover and **MW on High 5 minutes.** Then, **MW on 50% (Medium) 25 to 30 minutes,** stirring once midway through cooking. Let stand, covered, 10 minutes before serving. Makes 6 servings.

Serving Suggestion: *Beef Bourguignonne is traditionally served over cooked noodles.*

VEAL BELLEVOIR FROM ZURICH

   ¼ cup butter
   3 green onions with tops, thinly sliced
   8 ounces fresh mushrooms, sliced
   1 pound boneless veal, julienned
   ¼ cup flour
   ½ teaspoon salt
   ¼ teaspoon white pepper
   ½ cup dry white wine
   ¾ cup beef broth or bouillon
   1 cup heavy cream
   2 teaspoons Kitchen Bouquet

Place butter, onions and mushrooms in a 2-quart casserole. Cover and **MW on High 3 minutes.** Add veal, re-cover, and **MW on High 3 minutes.** Using a slotted spoon, remove veal and mushrooms; set aside. Blend flour, salt and pepper into butter and drippings. Using a wire whisk, blend in wine, bouillon, cream and Kitchen Bouquet. Whisking several times, **MW on 50% (Medium) 8 minutes,** or until thickened. Add reserved veal and mushrooms to thickened sauce. Reheat on 50% to desired temperature. Makes 4 servings.

Serving Suggestion: *Sprinkle with snipped fresh parsley to serve. The classic accompaniment is "Kartöffelrosti" (see page 145).*

LIVER

Due to the fat distribution in liver, it sometimes makes a "popping" sound when microwaved. This doesn't hurt anything—just keep the utensil covered to prevent oven spatters.

Liver freezes well for six months. To **defrost** in the microwave oven, leave in original carton. **MW on 30% (Medium-low) 5 to 7 minutes per pound.** Remove from carton midway through defrosting. After defrost time, separate pieces of liver and rinse in cold water.

LIVER MARENGO

- 1 pound beef or calf liver
- 1 (1.5-ounce) packet spaghetti sauce mix
- ¼ cup flour
- 1 (16-ounce) can tomatoes with liquid
- ½ cup beef broth
- ¼ cup white wine

Slice liver into ½-inch wide strips. Place in a 2-quart casserole. Combine sauce mix and flour and add to liver. Toss to coat liver strips well. Cover and **MW on High 3 minutes.** Stir to redistribute liver strips. Re-cover and **MW on High 2 to 3 minutes.** Coarsely mash tomatoes or chop in food processor. Add tomatoes, beef broth and wine to liver; stir well. **MW on High 4 minutes,** or until heated through and thickened. Makes 4 servings.

Serving Suggestion: *Serve over cooked noodles or rice.*

LIVER CUTLETS

- 6 slices bacon
- ¾ cup finely-crushed buttery crackers (such as Ritz)
- 1 teaspoon onion salt or ½ teaspoon onion powder
- 1 pound beef or calf liver
- ⅓ cup Thousand Island dressing
- Paprika

Lay strips of bacon on bottom of a 2-quart rectangular dish. Cover with wax paper. **MW on High 6 minutes.** Remove bacon to paper toweling. Combine cracker crumbs with onion salt. Rinse liver slices and pat dry with paper toweling. Spread both sides of liver with dressing and dredge in crumb mixture. Place liver slices in bacon drippings in same dish. Cover with wax paper. **MW on High 3 minutes.** Turn slices of liver over and sprinkle with paprika. Re-cover and **MW on High 2 to 3 minutes.** Crumble cooked bacon over liver to serve. Makes 4 to 5 servings.

BACON

Bacon—it's one of the most popular foods to prepare in the microwave oven.

Regular sliced (as opposed to thick-sliced) bacon seems to microwave best. The brown spots and crustiness that appear on paper toweling and bacon racks are a result of the sugar used in the curing of the bacon. Canadian-style bacon is already pre-cooked and needs only to be heated before eating.

Several tips to keep in mind for microwaving bacon:

- To **defrost frozen bacon,** place unopened package of bacon in microwave oven and **MW on 30% (Medium-low) 4 to 5 minutes,** turning package over midway through defrosting time.
- To **separate slices** of refrigerated bacon more easily, microwave a few seconds. **One pound of bacon** will take approximately **30 seconds on High** to soften the pieces for easy separation.
- *Never* use newspapers in the microwave oven to absorb grease. The lead in the printer's ink combined with the hot grease could start a fire and yield some mighty hot news!
- A few strips of bacon can be microwaved on layered paper toweling placed on a paper plate. However, microwaving more than 5 pieces of bacon on a paper plate results in too much grease for this method. Use a bacon or meat rack instead.
- Bacon racks will clean more easily if you place a paper towel on the rack first. This will prevent the crusty sugar from sticking to the rack. Cover bacon completely with paper toweling to prevent splattering. Rotate bacon rack once or twice during microwaving for more uniform cooking.
- Diced bacon will cook faster if arranged in a wreath fashion so that the microwave energy can penetrate from the inner as well as the outer edges of the arranged bacon.
- Bacon that is diced before cooking will tend to clump together unless stirred once or twice during microwaving.
- Raw bacon added to other ingredients in a recipe (such as in Baked Beans) will remain very fatty and undesirable in texture and color. We recommend microwaving it before adding it to a recipe.
- Remove paper toweling and transfer cooked bacon to a plate. Bacon will continue to cook after microwaving and become crisp upon standing a few minutes. **Bacon is always microwaved on High power.** The following chart will serve as a guide for timing. However, thickness of bacon, amount of sugar and salt used in the curing process, brand of bacon, starting temperature, and amount of doneness desired will affect the cooking time.

BACON MICROWAVING CHART

1 slice - 1 minute	4 slices - 3 to 3¼ minutes
2 slices - 1¾ to 2 minutes	5 slices - 3¾ to 4 minutes
3 slices - 2¼ to 2½ minutes	6 slices - 4½ to 4¾ minutes

There are several methods that can be used for cooking bacon in your microwaving oven. Choose the method that suits your needs best.

STRIPS OF BACON

Method 1 (if grease is to be discarded): Place several layers of paper toweling on a paper plate and lay a single layer of separated pieces of bacon on top of toweling. Bacon pieces may need to be cut in half lengthwise so that they fit on the paper plate. Cover bacon with paper toweling and microwave, rotating plate once or twice. Dispose of paper plate and toweling.

Method 2 (if grease is to be saved): Place a paper towel on a bacon or meat rack. Lay a single layer of separated pieces of bacon on top of toweling. If desired, cover with two layers of paper toweling and lay crosswise another single layer of separated pieces of bacon on top. Cover bacon with paper toweling and microwave, rotating rack once or twice. Remove bacon and pour grease into container.

DICED BACON

Arrange uncooked diced bacon in a wreath fashion on several layers of paper toweling on a paper plate or directly in a casserole dish. Cover bacon with paper toweling and microwave. To prevent pieces of bacon from sticking to one another, stir once or twice during microwaving to redistribute pieces. Remove bacon pieces, discard paper toweling and grease.

If bacon grease is to be kept for future use, arrange diced bacon in wreath fashion directly in casserole dish. Proceed with above directions. Remove bacon and pour grease into container. See index for Bacon recipes.

FASTER THAN A SPEEDING BULLET

- 1 pound cabbage, cut into (1-inch) squares
- 1 teaspoon caraway seeds
- 1 pound fully-cooked link sausage
- 1 tablespoon margarine

Evenly distribute cabbage in a 2-quart rectangular glass dish. Sprinkle cabbage with caraway seeds. Cut sausage into (2-inch) chunks and place on top of cabbage. Cover dish tightly with plastic wrap. **MW on High 9 to 10 minutes,** rotating dish if necessary. Drain liquid and dot with margarine. Re-cover and let stand 3 minutes. Makes 4 servings.

Sausage ←

SAUSAGE RAGOUT

- 1 medium onion, chopped
- ¼ teaspoon instant minced garlic
- 1 pound fully-cooked link sausage, cut into 1-inch chunks
- 1 (16-ounce) can sliced stewed tomatoes, including liquid
- 1 (4-ounce) can mushroom pieces, drained
- 2 cups water
- 2 teaspoons instant beef bouillon granules
- 2¼ cups instant rice
- 1 (10-ounce) package frozen peas, thawed
- 1 cup shredded lettuce

Place onion and garlic into a 4-quart simmer pot. Cover and **MW on High 3 minutes.** Add sausage, tomatoes, mushrooms, water and bouillon. Cover and **MW on High 5 minutes.**

Stir in rice, cover and **MW on High 6 to 8 minutes,** or until most of liquid has been absorbed by rice. Stir in peas and lettuce, cover and let stand 5 minutes. *Freezes well: do not add lettuce before freezing.* Makes 6 to 8 servings.

FREEZER SAUSAGE CASSEROLES

- 2 pounds bulk sausage
- 3 cups instant rice
- 1 (4-serving size) packet dry onion soup mix
- 3½ cups boiling water
- 1 (10-ounce) package frozen cut green beans
- 2 cups shredded cheddar cheese (8 ounces)

Crumble sausage into a hard-plastic colander set in a 4-quart casserole. Cover and **MW on High 9 to 10 minutes,** stirring midway through cooking and at end. Discard grease and set sausage aside.

In same casserole, place rice and soup mix. Pour boiling water over them and cover; let stand 5 minutes. Meanwhile, place package of frozen beans on a paper towel and **MW on High 5 minutes.** Combine beans and sausage with rice mixture. Reserve ½ cup of cheese and stir remainder into rice mixture.

To serve immediately, divide mixture into two (2-quart) casseroles. For each casserole, **MW on High 4 to 5 minutes.** Sprinkle with ¼ cup shredded cheese and **MW on High 1 minute,** or until cheese melts. Each casserole makes 4 to 5 servings.

NOTE: *To freeze one or both casseroles, refer to Casserole Freezing Directions, page 101.*

Ham

NOW-AND-LATER HAM CASSEROLES

- 2 pounds zucchini, sliced
- 1 cup chopped onions
- 3 cups cooked rice
- 1 (4-ounce) can chopped green chilies, including liquid
- 1 pound cooked ham, ground (4 cups)
- ½ teaspoon garlic powder
- ¼ teaspoon pepper
- 2 eggs, beaten
- 12 ounces cottage cheese (1½ cups)
- 2 cups shredded cheddar cheese (8 ounces)

Combine zucchini and onions in a 2-quart batter bowl. Cover tightly with plastic wrap. **MW on High 8 to 10 minutes,** stirring once midway through cooking. In large mixing bowl, combine rice, chilies, ham, garlic powder, pepper, eggs and cottage cheese. Gently fold in zucchini and onions.

Mixture makes two (2-quart) rectangular casseroles. If you want to freeze one or both, line empty utensil(s) with aluminum foil and follow Casserole Freezing Directions, below. To serve one casserole immediately, pour half of mixture into a (2-quart) rectangular glass dish. Shield corners with aluminum foil (see page 183). Cover with plastic wrap and **MW on 70% (Medium-high) 10 minutes,** rotating dish midway through cooking. Remove plastic wrap and sprinkle with 1 cup shredded cheese. **MW on 70% until cheese melts.** Each casserole makes 6 to 8 servings.

CASSEROLE FREEZING DIRECTIONS

To prepare casserole for freezing, line two (1½-quart) round utensils with aluminum foil. Divide recipe into each. Cover with foil and crimp to foil lining. Label contents. If a cheese or topping is reserved to sprinkle on reheated casserole, place it in a small plastic bag and tape bag to foil covering. Freeze food. When frozen solid, remove utensil and return to kitchen use.

To defrost and reheat one frozen casserole, leave foil around sides of frozen food. Score with knife and pull off foil from top and bottom of frozen food. Place food in original utensil and **MW on 70% (Medium-high) 15 minutes,** stirring when possible. Sprinkle with reserved cheese or topping and **MW on 70% to serving temperature.**

Ham

HAM PINEAPPLE STACK-UPS

- 3 cups ground cooked ham
- ½ pound ground beef
- ½ cup cracker crumbs
- ¼ cup finely-chopped onion
- ¼ cup milk
- 1 egg
- 1 tablespoon parsley flakes
- ¼ teaspoon pepper
- 5 canned pineapple slices, reserve juice for glaze

Combine all ingredients except pineapple. Form mixture into 6 patties of equal size as the pineapple slices. Stack patties alternately with pineapple to form a loaf. Using two bamboo skewers, connect layers through pineapple slices. Place loaf on a microwave-safe tray and cover with wax paper. **MW on High 5 minutes.** Turn loaf upside down and glaze. **MW on High 3 to 5 minutes.** Decorate with cherries and baste with glaze. Re-cover with wax paper and let stand 5 minutes. Makes 5 servings.

GLAZE

- ½ cup packed dark brown sugar
- 2 tablespoons reserved pineapple juice
- 1 tablespoon cider vinegar
- 1 teaspoon prepared mustard
- 6 maraschino cherries

Combine ingredients to glaze ham loaf, and use cherries for decoration.

STUFFED SQUASH HALVES

- 1 butternut squash (about 1½ pounds)
- 2 cups finely-ground cooked ham
- 2 slices bread, finely crumbed
- 2 tablespoons pineapple preserves
- ⅛ teaspoon nutmeg

Pierce squash with a paring knife and **MW on High 3 minutes.** Slice squash in half lengthwise; remove seeds.

Combine ham, bread crumbs, preserves and nutmeg. Divide ham mixture between squash halves and mound on top. Place squash halves in a 1½-quart rectangular dish and cover with plastic wrap. **MW on High 10 to 12 minutes,** or until squash is almost done, rotating dish twice. Let stand 5 minutes. Makes 2 to 4 servings.

HAM 'N CHEESE PATTIES

- 1 pound ground cooked ham
- 4 slices bread, crumbed
- ½ cup shredded cheddar cheese
- ¼ cup chopped onion
- ¼ cup milk
- 2 eggs
- 2 tablespoons snipped fresh parsley
- 1 tablespoon prepared mustard
- ¼ teaspoon pepper
- ¼ cup cornflake crumbs

Combine all ingredients except cornflake crumbs; mix well. Divide mixture and form into 8 balls. Roll balls in cornflake crumbs and place on a microwave-safe tray. Flatten balls slightly into patties. Cover with wax paper and **MW on 70% (Medium-high) 12 to 14 minutes,** rotating tray midway through cooking. Let stand 5 minutes before serving. *Cooked patties freeze well for later use.* Makes 4 to 6 servings.

SANDY'S HAM STRATA

- 1 (10-ounce) package frozen chopped broccoli
- 1 (6-ounce) package seasoned croutons
- 1½ cups diced cooked ham
- 2 cups shredded sharp cheddar cheese (8 ounces)
- 4 eggs
- 1⅓ cups milk
- Paprika

Place package of broccoli in a 4-cup glass measure. **MW on High 5 minutes.** Discard liquid that has drained into measure. Set broccoli aside.

Distribute croutons in bottom of a 9-inch microwave-safe layer cake pan. Make one layer of each: ham, broccoli; then cheese.

Whisk eggs and milk together. Pour over layered mixture. Dust top with paprika. Cover and refrigerate overnight.

Remove cover and **MW on 70% (Medium-high) 15 to 17 minutes,** rotating dish twice. Let stand 5 minutes before cutting. Makes 6 to 8 servings.

Ham

CAROLINA HAM SLICES
In honor of our RALEIGH NEWS AND OBSERVER

- 2 (1-pound) center-cut ham slices
- 2 cups sliced apples
- ⅓ cup raisins
- ⅓ cup packed dark brown sugar
- ¼ cup dark corn syrup
- ½ teaspoon ginger
- Apple juice or water
- 2 teaspoons cornstarch

Remove excess fat and center bone from ham slices. Place one ham slice in a 1½-quart rectangular glass dish. Combine apples, raisins, sugar, syrup and ginger in a medium-size bowl. Distribute ⅔ of the apple mixture over top of ham slice in dish. Place remaining ham slice on top of apple mixture and distribute remaining apple mixture on top. Cover with plastic wrap and **MW on 50% (Medium) 18 to 20 minutes,** rotating dish twice.

Pour liquid into a 2-cup glass measure. If necessary, add apple juice to yield 1-cup liquid. Whisk in cornstarch. Whisking midway through cooking, **MW on High 2½ to 3 minutes,** or until thickened. (This is a light sauce – slightly thickened.) Serve sauce in a pitcher to accompany entree. Makes 5 to 6 servings.

BETTY'S BONANZA

- ⅔ cup chopped onion
- 2½ cups cooked rice
- 2¼ cups diced cooked meat (ham, chicken, etc.) or 1¼ pounds cooked shrimp
- 1 (10¾-ounce) can cream of mushroom soup
- ½ cup sour cream
- ¾ cup shredded cheddar cheese
- ½ teaspoon garlic powder
- Salt and pepper to taste

Place onion in a 2-quart round casserole. Cover with plastic wrap and **MW on High 3 to 3½ minutes.** Blend in rice, meat, soup, sour cream, ½ cup cheese, garlic powder, salt and pepper. Cover. Stirring midway through cooking, **MW on 70% (Medium-high) 6 to 8 minutes,** or until heated through. Garnish top with remaining cheese and **MW on 70% 1 to 1½ minutes** to melt cheese. Makes 6 servings.

NOTE: *Casserole may be assembled ahead of time and refrigerated.* **MW on 70% 9 to 12 minutes** *to heat through.*

Pork

BROCCOLI HAM AU GRATIN

- 1 (10-ounce) package frozen broccoli cuts
- 2 tablespoons margarine
- 2 tablespoons flour
- ½ teaspoon dry mustard
- 1 cup milk
- ¾ cup shredded sharp cheddar cheese (6 ounces)
- 1 pound fully cooked ham, cut in ½-inch cubes
- 4 to 5 large potatoes, baked

Place package of broccoli cuts on a paper plate. **MW on High 4 to 4½ minutes.** Drain and reserve.

Place margarine in a 1½-quart casserole. **MW on High 30 seconds,** or until melted. Stir in flour and dry mustard. Blend in milk with a wire whisk. Whisking midway through cooking, **MW on High 3 to 3½ minutes,** or until thickened. Blend in cheese until melted.

Add cooked broccoli and ham. **MW on 70% (Medium-high) 4 to 5 minutes,** or until heated through. *Serve as a potato topper.* Makes 4 to 5 servings.

STUFFING TOP CHOPS

1 (6-ounce) box stuffing mix, any flavor
1½ pounds center-cut pork chops (4 to 6)

Make stuffing according to package directions; set aside. If using a 10-inch browning dish, **preheat on High 6 minutes.** Coat surface quickly with 1 teaspoon margarine and press chops onto hot surface to brown. When meat stops sizzling, turn over and arrange with bone toward the center and meat to the outside of the dish. Divide stuffing among pork chops and mound on top. Cover with glass lid and **MW on 50% (Medium) 20 to 25 minutes,** rotating dish midway through cooking. Let stand 10 minutes. Serves 4 to 6.

NOTE: *Can also be microwaved in a 2-quart rectangular casserole covered with plastic wrap or glass lid. Timing is the same.*

CRUMB-COATED TENDERLOIN

1½ pounds boneless pork tenderloin
1 egg, beaten
Italian-flavored dry bread crumbs

Coat tenderloin in beaten egg and roll generously in bread crumbs. Place on a microwave meat or bacon rack. Cover with wax paper. **MW on 50% (Medium) 18 to 20 minutes,** turning top to bottom midway through cooking. *If using temperature probe, set for 170°.* Let stand 5 to 10 minutes before slicing. Serves 6.

Quick-Tip → If microwaving sausage patties, use a browning dish to achieve a traditional crusty brown exterior. Without a browning dish, use Kitchen Bouquet or Micro Shake to add appealing color.

POULTRY

Your microwave oven is the quickest and easiest way to prepare poultry. Skin poultry if desired and **MW on High allowing 6 to 7 minutes per pound.** Microwave individual chicken pieces on a meat or bacon rack covered with wax paper. A whole chicken or turkey can be microwaved in an oven cooking bag. After microwaving poultry, be sure to allow it to stand for one-third to one-half of the cooking time. For additional information and more poultry recipes, see our *Microwave Know-How* book, *MicroScope Savoir Faire.* (See coupon in back for ordering information.)

CRISPY BUTTERNUT CHICKEN

- ½ cup dry roasted peanuts
- ¼ cup cornflake crumbs
- ⅓ cup buttermilk-based salad dressing
- 2½ pounds chicken pieces or cut-up fryer
- Paprika

Place peanuts into blender or food processor. Chop fine. Combine on a piece of wax paper with cornflake crumbs. Pour salad dressing into a pie plate. Remove skin from chicken and dip each piece in salad dressing. Roll in crumb mixture to coat. Arrange chicken on a microwave meat or bacon rack. Place the thickest pieces toward the outside and the thin or bony pieces in the center. Sprinkle with paprika. Cover with wax paper and **MW on High 15 to 18 minutes,** rotating rack midway through cooking. Let stand 8 minutes. Makes 4 servings.

CHICKEN 'N STUFFIN'

- 1 (6-ounce) box Stove Top cornbread stuffing mix
- 2½ pounds chicken breast halves
- Oil, Kitchen Bouquet, paprika

Into a 4-cup glass measure, place water and margarine specified on box. Add contents of seasoning packet. **MW on High 3 minutes,** or until boiling. Place dry stuffing in a 2-quart rectangular glass dish. Add water mixture; stir and level. Arrange chicken on top of stuffing with meaty areas toward outside of dish. Brush with oil and Kitchen Bouquet; dust with paprika. Cover with wax paper and **MW on High 17 to 19 minutes,** rotating dish midway through cooking. Makes 4 to 6 servings.

Chicken

APPLE STUFFED CORNISH HENS

- ⅓ cup water
- 3 tablespoons margarine
- ¼ cup chopped celery
- 1 teaspoon instant minced onion
- 1 cup dry herb stuffing mix (such as Pepperidge Farm)
- ¼ cup dry roasted peanuts
- 1 tart cooking apple, cored and diced
- 2 (10-ounce) Cornish game hens, defrosted
- Oil
- Kitchen Bouquet

Combine water, margarine, celery and onion in a 4-cup glass measure. **MW on High 1½ minutes,** or until margarine melts and water is boiling. Stir in stuffing mix until liquid is absorbed. Stir in peanuts and apple.

Remove neck and giblets from hens. Wash in cold water and pat dry. Fill cavities with stuffing and truss to secure drumsticks. *If any stuffing will not fit into hens, place in baking dish along with hens to cook.* Place breast-side up in a 1½-quart rectangular dish. Brush with oil and Kitchen Bouquet. Garnish with paprika. **MW on 70% (Medium-high) 18 to 20 minutes,** turning each hen ½-turn around midway through cooking. Let stand 5 minutes before serving. Makes 2 to 4 servings.

TERIYAKI CHICKEN

- ½ cup soy sauce
- ¼ cup dry white wine
- Grated rind from one orange
- 1 tablespoon sugar
- ½ teaspoon ground ginger
- ½ teaspoon instant minced garlic
- 3 pounds chicken pieces
- Paprika

Combine soy sauce, wine, grated rind, sugar, ginger and garlic in a 2-quart rectangular dish. Wash and dry chicken pieces and place in marinade, turning to coat each piece. Cover and marinate 2 hours, or as long as overnight, basting with marinade occasionally.

To cook, pour off half of marinade and arrange chicken skin-side down, with the meaty portions to the outside of the dish and the bony pieces in the center. Cover with wax paper and **MW on High 9 minutes.** Turn each piece over, baste with marinade and sprinkle with paprika. Re-cover and **MW on High 9 to 11 minutes.** Let stand 5 minutes before serving. Makes 4 to 5 servings.

▶ *Chicken*

CHICKEN HUNGARIAN

1 large onion, chopped
3 tablespoons margarine
2 tablespoons tomato paste
1½ tablespoons paprika
½ cup chicken broth
3 pounds chicken pieces, skin removed
1 cup sour cream

Place onion and margarine in a 4-cup glass measure. Cover with plastic wrap and **MW on High 4 to 5 minutes.** Stir in tomato paste and paprika. Blend in broth. Place chicken in a 2-quart rectangular dish. Pour onion mixture over chicken and turn to coat. Arrange chicken with the meaty portions to the outside of the dish and the bony pieces in the center. Cover with wax paper and **MW on High 9 minutes.** Turn each piece over and baste with sauce. Re-cover and **MW on High 9 to 11 minutes.** Stir sour cream into chicken and sauce. Cover and let stand 5 minutes. If necessary to reheat, **MW on 70% (Medium-high).** Makes 5 to 6 servings.

Serving Suggestion: *Place chicken on a bed of cooked egg noodles and spoon sauce over both.*

HONEYED CHICKEN

3 tablespoons margarine
⅓ cup honey
2 tablespoons prepared mustard
¾ teaspoon curry powder
½ teaspoon salt
3 pounds chicken pieces
Paprika

Place margarine in a 2-quart rectangular dish and **MW on High 40 seconds.** Stir in honey, mustard, curry powder and salt. Roll chicken pieces in mixture and arrange in a single layer in dish, with meaty portions to the outside of the dish and the bony pieces in the center. Cover with wax paper and **MW on High 9 minutes.** Turn each piece over, baste with sauce, and sprinkle with paprika. Re-cover and **MW on High 9 to 11 minutes.** Let stand 5 minutes. Makes 4 to 5 servings.

Serving Suggestion: *Serve with curried rice.*

Chicken

CHICKEN CUTLETS MOZZARELLA

 1 pound boneless chicken breasts
 ½ cup Italian-seasoned dry bread crumbs
 1 cup prepared meatless spaghetti sauce
 4 slices mozzarella cheese

Separate chicken breasts into four portions and pound to flatten slightly. Coat well in bread crumbs. Arrange breasts around sides of an 8-inch baking dish. Cover with wax paper. **MW on High 5 to 6 minutes.** Pour one-fourth of the sauce over each portion. Place a slice of cheese on top of each and garnish with crumbs, if desired. Cover with wax paper and **MW on 70% (Medium-high) 3 to 4 minutes,** or until sauce is hot and cheese is melted. Makes 4 servings.

SUPREME ROLLS

 4 boneless chicken breasts, skinned and halved
 8 (thin) slices fully-cooked ham
 1 (6-ounce) package sliced Swiss cheese
 2 tablespoons margarine
 2 tablespoons flour
 1 cup chicken stock
 1 (4-ounce) can mushroom pieces, drained
 2 tablespoons snipped fresh parsley

Place one half breast on top of each piece of ham. Divide cheese and place over top of chicken. Roll up each ham-chicken-cheese packet and place seam-side down in a 2-quart rectangular dish. Cover with wax paper and **MW on High 6 minutes.** Using tongs, redistribute rolls. Re-cover and **MW on High 6 minutes.** Let stand 5 minutes.

Place margarine in a 4-cup glass measure. **MW on High 30 seconds.** Blend in flour, then chicken stock. Stirring once, **MW on High 3 to 4 minutes,** or until thickened. Add mushrooms and parsley. **MW 1 to 1½ minutes,** or until mushrooms are heated. *Serve with Spinach Fettuccine, page 62.* Makes 6 to 8 servings.

Quick-Tip ➡ Deboning poultry is much easier when done while it is still warm. If you can bring yourself to face the big bird after dinner, remove meat and refrigerate. Use leftover meat in any recipe requiring cooked chicken or turkey.

MACADAMIA NUT CHICKEN

1½ pounds boneless chicken breasts, skinned
¼ cup margarine
1¼ cups finely-crushed Ritz crackers
½ teaspoon ginger
⅛ teaspoon garlic powder
1½ cups pineapple juice
2 tablespoons cider vinegar
2 teaspoons soy sauce
¼ cup packed brown sugar
3 tablespoons cornstarch
½ cup chopped Macadamia nuts

Cut chicken into bite-size pieces. Place margarine in a 2-quart rectangular dish and **MW on High 1 minute.** Add chicken pieces and coat with margarine. Combine cracker crumbs, ginger and garlic powder. Pour over chicken, stirring to completely cover chicken with crumbs. Cover with wax paper and **MW on High 6 to 7 minutes,** redistributing pieces midway through cooking. Let stand while making sauce.

Combine pineapple juice, vinegar and soy sauce in a 4-cup glass measure. Mix sugar and cornstarch together. Using a whisk, blend into liquid mixture. Whisking midway through cooking, **MW on High 3 minutes,** or until thickened. *To serve, place chicken on bed of regular or Fried Rice, page 154.* Pour sauce over top. Garnish with nuts. Makes 4 to 5 servings.

GINGERED CHICKEN

1¼ pounds chicken breasts, skinned and boned
1½ tablespoons margarine
2 teaspoons cornstarch
1 (4-ounce) can mushroom pieces, including liquid
1 (8-ounce) can sliced water chestnuts, drained
1 teaspoon ginger
2 teaspoons soy sauce
3 green onions with tops, thinly sliced

Cut chicken into bite-size pieces. Combine with margarine in a 1½-quart casserole. Cover and **MW on High 2 to 2½ minutes.** Stir in cornstarch and mix thoroughly. Add mushrooms, including liquid and water chestnuts; stir to combine. Blend ginger into soy sauce and stir into meat mixture. Cover and **MW on High 5 to 6 minutes,** stirring midway through cooking. Add green onions, re-cover and **MW on High 2 to 2½ minutes.** Let stand 5 minutes. Makes 4 to 5 servings.

CHICKEN CASHEW

- 1 pound boneless chicken breasts, cut into bite-size pieces
- ½ cup white wine
- 3 tablespoons soy sauce
- 1 tablespoon cornstarch
- ½ teaspoon instant minced garlic
- ¼ teaspoon ginger
- 1 medium green bell pepper, cut into bite-size pieces
- ½ cup cashew halves

Place chicken pieces into a 2-quart round casserole. Combine wine, soy sauce, cornstarch, garlic and ginger in a 1-cup glass measure. Pour over chicken. Cover with wax paper and **MW on High 4 minutes,** stirring midway through cooking. Add bell pepper; re-cover. **MW on High 3 to 4 minutes,** stirring midway through cooking. Let stand 3 minutes. Serves 4 to 5.

CHICKEN & NOODLES PAPRIKA

- 1 cup chopped onion
- 2 tablespoons margarine
- 2 tablespoons flour
- 1 tablespoon paprika
- 1 teaspoon basil
- ½ teaspoon salt
- ⅛ teaspoon pepper
- 1 cup chicken broth
- 2 tablespoons dry white wine, optional
- ½ cup plain yogurt
- 1 pound cooked chicken, shredded or finely diced
- 2 cups cooked noodles

Combine onion and margarine in a 1½-quart round casserole. Cover and **MW on High 3 to 3½ minutes.** Stir in flour, paprika, basil, salt and pepper. Blend in chicken broth using a whisk. Stirring midway through cooking, **MW on High 3½ to 4 minutes,** or until thickened. Stir in wine and yogurt. Add chicken and noodles; stir to combine. **MW on 70% (Medium-high) 6 to 7 minutes,** or until heated through. Makes 4 to 5 servings.

> Turkey

TURKEY TENDERLOIN VINAIGRETTE

1 pound fresh turkey tenderloin
¼ cup dry white wine

There are usually two turkey tenderloins in a 1-pound package. Arrange them in a 9-inch round cake or pie pan, with ends overlapping making a wreath. Pour wine over tenderloins and cover with plastic wrap. **MW on High 5 to 6 minutes,** rotating midway through cooking. Let stand until barely warm. Pour off wine and refrigerate. Slice and serve with Vinaigrette Sauce, below. Makes 4 to 5 servings.

VINAIGRETTE SAUCE

4 tablespoons wine vinegar
½ cup olive oil
1 tablespoon dried parsley flakes
1 tablespoon chopped green onion
½ teaspoon dry mustard
¼ teaspoon basil
Pinch of tarragon
Pinch of garlic powder
Salt and pepper
1 hard-cooked egg, chopped

Combine all ingredients and let stand at room temperature several hours for flavors to blend. *Olive oil solidifies when refrigerated.* Yields 1 cup.

ASPARAGUS TURKEY SPAGHETTI

6 ounces spaghetti, cooked
1 (10½-ounce) can cream of asparagus soup
⅔ cup milk
¼ teaspoon basil
⅓ cup grated parmesan cheese
2 cups chopped cooked turkey
1 (10-ounce) can cut asparagus, drained
¼ cup slivered toasted almonds

In a 2-quart round casserole, combine spaghetti, soup, milk and basil. Reserve 2 tablespoons cheese for topping casserole. Add remaining cheese, turkey and asparagus to spaghetti mixture. Sprinkle with reserved cheese and almonds. **MW on 70% (Medium-high) 6 to 7 minutes,** or until bubbly hot. Makes 4 servings.

TURKEY HURRY CURRY

½ cup chopped onion
½ cup chopped celery
½ cup margarine
½ teaspoon salt
1 tablespoon curry powder
½ cup flour
2 cups turkey or chicken stock
1 cup milk
3 cups chopped cooked turkey
2 tablespoons dry sherry

Place onion, celery and margarine in a 3-quart round casserole. **MW on High 3 minutes.** Using a whisk, blend salt, curry powder and flour into margarine mixture. Whisk in stock and milk. Whisking every 3 minutes, **MW on High 8 minutes,** or until thickened. Add turkey and sherry. **MW on 70% (Medium-high) 5 minutes.** Makes 8 servings.

Serving Suggestion: *Serve over a bed of hot rice with an assortment of condiments, such as chopped peanuts, chopped hard-cooked eggs, diced cooked bacon, flaked coconut, chow mein noodles, chopped olives, raisins or chutney purchased in jar.*

CHINESE TURKEY

¼ cup margarine
1 cup thinly-sliced celery
¼ cup flour
½ teaspoon ginger
¼ teaspoon instant minced garlic
1 tablespoon soy sauce
2 cups chicken broth
2 cups diced cooked turkey or chicken
1 (6-ounce) package frozen pea pods, defrosted
1 (5-ounce) can Chinese chow mein noodles

Place margarine and celery in a 2-quart casserole. Cover and **MW on High 5 minutes.** Blend in flour, ginger, and garlic. Using a whisk, gradually stir in the soy sauce and broth. Whisking every two minutes, **MW on High 4 minutes,** or until sauce is thickened. Stir in turkey and pea pods. **MW on High 4 minutes,** or until heated through. Serve over noodles. Makes 4 servings.

DEFROSTING FISH

To thaw a 1-pound carton of frozen fish, crimp small strips of aluminum foil to the ends of the carton. This will prevent the fish from cooking on the ends during the defrosting process. For a 1-pound carton, **MW on 30% (Medium-low) 8 minutes,** flipping the box over midway through defrosting. Home-frozen fish usually takes about 6 minutes per pound to defrost.

Instead of separating the icy cold fillets into individual fillets, the Canadian Fish Council suggests cutting directly across the block of fillets into the desired number of serving portions. The uniform size and shape of these portions of fish microwave well. After cutting, allow a few minutes for the fish to finish thawing before microwaving.

CHIPPER FISH

- 1 (1-pound) carton frozen Canadian Cod, defrosted
- ⅓ cup Caesar salad dressing
- 1 cup crushed potato chips

Cut directly across the block of fillets into four equal portions. Coat fillets with salad dressing. Place fish in a 9-inch glass pie plate. Sprinkle potato chips over fish. So that chips will stay crisp, do not cover. **MW on High 4 to 5 minutes,** rotating dish midway through cooking. Let stand 3 minutes, or until fish flakes easily before serving. *Serve with a wedge of lemon.* Makes 4 servings.

EASY CHEESY COD

- 1 (1-pound) carton frozen Canadian Cod, defrosted
- 2 tablespoons lemon juice
- 3 green onions with tops, thinly sliced
- ¾ cup shredded cheddar cheese
- 1 medium tomato, sliced horizontally into fourths
- Salt and pepper

Cut directly across the block of fillets into four equal portions. Place fish in a 9-inch glass pie plate. Drizzle lemon juice over top of fish. Distribute onions on top of fish. Sprinkle ½ cup of cheese over top of green onions. Place tomato slices on top of cheese. Cover with wax paper. **MW on High 4 to 5 minutes,** rotating dish midway through cooking. Sprinkle remaining cheese on top, re-cover and let stand 3 minutes, or until fish flakes easily before serving. Season to taste. Makes 4 servings.

SPINACH STUFFED FILLETS

½ cup chopped onion
2 tablespoons margarine
1 (10-ounce) box frozen chopped spinach
1 cup dry herb stuffing
1 pound fresh or defrosted frozen perch fillets
Fresh lemon
¼ cup dry white wine

Combine onion and margarine in a 1-quart casserole. Cover and **MW on High 2 to 2½ minutes.** Unwrap spinach and place in same casserole. **MW on High 5 to 6 minutes,** breaking up frozen spinach with a fork midway through cooking. Do not drain. Add stuffing and stir; set aside.

Place fillets skin-side down, and sprinkle with fresh lemon juice. Divide spinach stuffing equally among fillets. Overlap ends of each fillet and secure with toothpicks.

Stand up stuffed fillets in a microwave ring pan. Pour wine around fish. Cover with wax paper and **MW on High 4 to 5 minutes,** rotating pan midway through cooking. Makes 4 servings.

RED SNAPPER RING-A-ROUND

In honor of our BEAUMONT ENTERPRISE

1 small onion
1 small green bell pepper
3 tablespoons butter
1 tablespoon lemon juice
1 pound Red Snapper fillets
Paprika

Thinly slice onion and pepper and separate into rings. Set aside.

Place butter in a 9-inch glass pie plate and **MW on High 1 minute,** or until melted. Stir in lemon juice. Arrange onion rings around outside edge of dish, leaving center area open. Place fillets in a circle on top of onion rings, tucking thin edge of fillet beneath thicker portion of next fillet. Tilt dish to spoon lemon butter over top of fish. Arrange pepper rings on top of fish. Dust with paprika. Cover with wax paper and **MW on High 4½ to 5½ minutes,** rotating dish midway through cooking. Let stand covered 3 minutes before serving. Makes 3 to 4 servings.

Salmon

CURRIED SALMON RING

- 1 (15½-ounce) can salmon
- 2 slices whole wheat bread
- 2 eggs
- ¼ cup milk
- 2 tablespoons finely-minced onion
- 2 tablespoons snipped fresh parsley
- ½ teaspoon curry powder
- ¼ teaspoon pepper

Drain salmon; remove bones and skin. Place in mixing bowl. Using food processor or blender, crumb bread. Add to salmon, along with remaining ingredients. Combine and pack into a 1-quart microwave-safe ring mold. Cover with wax paper and **MW on 50% (Medium) 12 to 14 minutes,** rotating midway through cooking. Let stand covered 6 minutes. Unmold on plate and top with Curry Sauce, see page 136. Makes 4 to 6 servings.

WINE-POACHED SALMON STEAKS

In honor of our Salem newspaper, the STATESMAN JOURNAL

- ½ cup white wine
- 2 (8-ounce) salmon steaks
- 4 tablespoons butter
- 1 clove garlic, minced
- 1 tablespoon lemon juice
- 2 tablespoons snipped fresh parsley
- ¼ teaspoon crushed marjoram or rosemary
- Salt and pepper to taste

Place wine in a 9-inch glass pie plate and **MW on High 2½ to 3 minutes,** or until boiling. Carefully place salmon steaks in wine and cover with plastic wrap. **MW on High 2½ to 3 minutes.** Turn fish over; re-cover. Let stand 5 minutes, or until fish flakes easily, before serving.

Place butter and garlic in a 1-cup glass measure. **MW on High 2 to 2½ minutes,** stirring twice. Blend in lemon juice and remaining seasonings.

Remove salmon from wine and drizzle herbed butter over top to serve. Makes 2 to 3 servings.

Quick-Tip ➡ Reheating fish toughens it, so microwave just before serving.

Shrimp

CHEDDAR SHRIMP CASSEROLE

In honor of our JACKSONVILLE JOURNAL

- 1 (5⅓-ounce) can evaporated milk
- ¼ cup ketchup
- 1 teaspoon worcestershire sauce
- 2 tablespoons frozen chopped chives
- 1 cup instant rice
- 1 (4-ounce) can mushroom pieces, drained
- 1 (10-ounce) package frozen cooked shrimp
- 1½ cups shredded mild cheddar cheese (6 ounces)

In a 2-quart round casserole, combine milk, ketchup, worcestershire sauce and chives. Stir in rice, mushrooms, frozen shrimp and cheese. Cover and **MW on High 5 minutes.** Stir well, re-cover and **MW on 50% (Medium) 10 to 12 minutes.** Stir, re-cover and let stand 5 minutes for rice to absorb liquid. *If desired, dust with paprika before serving.* Makes 4 to 5 servings.

SHRIMP WITH CASHEWS

- 2 tablespoons oil
- 3 green onions with tops, thinly sliced
- 1 pound raw medium shrimp in shells, peeled and deveined
- ½ (8-ounce) can sliced water chestnuts
- 2 teaspoons cornstarch
- 1 tablespoon sherry
- ¼ cup chicken broth
- 2 tablespoons soy sauce
- 1 teaspoon ginger
- ¼ teaspoon white pepper
- 1 (6-ounce) package frozen pea pods, defrosted
- ½ cup roasted cashews

Combine oil, onions and shrimp in a 2-quart round casserole. Cover and **MW on High 3 minutes.** Add water chestnuts and stir to redistribute shrimp. Cover and **MW on High 2 minutes.**

In a small bowl, combine cornstarch with sherry until smooth. Add broth, soy sauce, ginger and pepper. Add sauce to shrimp; stir. **MW on High 2 minutes.** Add pea pods. Cover and **MW on High 2 minutes,** or until heated through and sauce is thickened. Top with cashews and serve. Makes 4 servings.

Shellfish

SHRIMP SAHIB

- 2 tablespoons margarine
- ¼ cup chopped green bell pepper
- ¼ cup chopped onion
- 1 pound raw medium shrimp in shells, peeled and deveined
- 1 (10¾-ounce) can condensed cream of shrimp soup
- 1 (4-ounce) can mushroom pieces, drained
- 2 tablespoons white wine
- 1 teaspoon curry powder

Combine margarine, green pepper and onion in a 2-quart round casserole. Cover and **MW on High 2 minutes.** Add shrimp, cover and **MW on High 3 minutes.** Blend in soup, mushrooms, wine and curry powder. **MW on High 3 minutes.** Makes 4 servings.

Serving Suggestion: *Serve over rice, biscuits, or toast. Garnish with chopped peanuts and shredded coconut, or other favorite curry condiments.*

SCALLOPS BONAVENTURE

- 1 small green bell pepper, sliced
- 3 tablespoons margarine
- 1 pound bay scallops
- 2 tablespoons dry sherry
- 4 tablespoons flour
- 2 green onions with tops, thinly sliced
- 1 cup milk
- Salt, pepper and cayenne powder to taste
- Cooked rice or patty shells

Place bell pepper and margarine in a 2-quart glass bowl. Cover with plastic wrap and **MW on High 3 to 4 minutes.** Remove bell pepper using a slotted spoon. Reserve.

Drain liquor from scallops, reserving ¼ cup. Add scallops and sherry to margarine in bowl. Re-cover and **MW on High 2 minutes,** stirring after 1 minute. Remove scallops with a slotted spoon. Reserve.

Add flour and green onions to liquid in bowl; whisk. Blend in milk and reserved scallop liquor. Whisking twice during cooking, **MW on High 3½ to 4 minutes,** or until mixture thickens. Add reserved bell pepper and scallops. Season as desired with salt, pepper and cayenne pepper. **MW on 70% 1 to 2 minutes,** or to desired serving temperature. Serve over cooked rice or patty shells. Makes 4 to 5 servings.

Shellfish ←

ANNAPOLIS STUFFED CRAB

In honor of our JOURNAL NEWSPAPERS in suburban Washington, D.C.

- 4 tablespoons butter
- 1 cup finely-chopped onions
- 2 ribs celery, chopped
- 1 pound crabmeat
- ½ cup light cream
- 4 tablespoons chili sauce
- 2 teaspoons lemon juice
- 1 teaspoon worcestershire sauce
- 6 dashes Tabasco
- Salt and pepper to taste
- 1 cup crushed cracker crumbs

Place butter, onions and celery in a 2-quart glass bowl. Cover and **MW on High 5 minutes.** Add crabmeat, cream, chili sauce, lemon juice, worcestershire sauce and Tabasco; mix well. Add salt and pepper to taste.

Divide mixture into four scallop or crab shells. Top with cracker crumbs. If desired, drizzle with a little melted butter and dust with paprika. Place stuffed shells on a microwave-safe tray and **MW on 70% (Medium-high) 8 to 10 minutes,** or until heated through. *Usually served with a lemon wedge.* Makes 4 servings.

NOTE: *To heat one shell,* **MW on 70% 2 to 3 minutes.**

PAELLA PRONTO

- 1 green bell pepper, chopped
- 1 cup chopped onion
- 1 clove garlic, minced
- 2 tablespoons bacon fat or oil
- 1 (16-ounce) can sliced stewed tomatoes, undrained
- 6 dashes Tabasco sauce
- ⅛ teaspoon cayenne pepper
- 1 (4-ounce) can mushroom pieces, drained
- ½ pound raw shrimp, peeled and deveined
- 1 (6½-ounce) can clams, drained
- 1 cup diced cooked chicken
- 2 cups hot cooked rice

Combine green pepper, onion, garlic and bacon fat in a 1½-quart round casserole. Cover and **MW on High 2 minutes,** or until onion is limp. Add remaining ingredients except for rice. **MW on High 6 to 7 minutes,** or until shrimp are pink, stirring midway through cooking. Serve over hot cooked rice. Makes 4 servings.

Tuna

CASHEW TUNA CASSEROLES

- 2 (10¾-ounce) cans cream of mushroom soup
- 1 cup milk
- 2 (6½-ounce) cans tuna, drained
- 2 cups thinly-sliced celery
- 10 ounces chow mein noodles
- 1 (6-ounce) jar dry-roasted cashews

In a large mixing bowl, blend together soup and milk. Stir in tuna and celery. Reserve 1 cup of noodles for top of casseroles. Split each cashew in half. Fold cashews and remaining noodles into tuna mixture.

To serve immediately, divide mixture into two (1½-quart) round dishes. **MW each casserole on High 4 to 5 minutes.** Sprinkle top with ½ cup reserved noodles and **MW on High** until heated through. Each casserole makes 4 to 5 servings.

NOTE: *To freeze one or both casseroles, line empty utensils with aluminum foil and follow Casserole Freezing Directions, page 101.*

TUNA-NOODLE RING

- 2 tablespoons margarine
- ½ cup chopped green bell pepper
- ½ cup chopped celery
- 3 green onions with tops, thinly sliced
- 2 tablespoons flour
- 1 cup milk
- 1¼ cups shredded Monterey Jack cheese
- 8 ounces noodles, cooked and drained
- 2 (6½-ounce) cans tuna, drained
- 1 (4-ounce) can mushroom pieces, drained

Place margarine, pepper, celery and onions in a 2-quart batter bowl. Cover and **MW on High 3 to 3½ minutes.** Stir in flour. Blend in milk with a whisk. Whisking midway through cooking, **MW on High 2½ to 3 minutes,** or until thickened. Stir in cheese until melted. Add noodles, tuna and mushrooms; blend.

Pour mixture into a ring-mold utensil that has been sprayed with Pam and cover with wax paper. Rotating dish midway through cooking, **MW on 70% (Medium-high) 5 to 6 minutes,** or until heated through. Let stand 3 minutes. Invert on serving platter. *Center area may be filled with cooked green vegetable, if desired.* Makes 6 servings.

TUNA TOPPER

- 4 tablespoons margarine
- 4 tablespoons flour
- Dash of ground rosemary
- 1 (14½-ounce) can chicken broth
- 1 (6½-ounce) can tuna, drained
- 1 (4-ounce) can mushroom pieces, drained
- ½ cup sliced ripe olives
- Parmesan cheese

Place margarine in a 1½-quart casserole. **MW on High 45 to 60 seconds,** or until melted. Stir in flour and rosemary. Blend in chicken broth with a whisk. Whisking midway through cooking, **MW on High 5 to 5½ minutes,** or until thickened. Add tuna, mushrooms and ripe olives. **MW on High 2 to 3 minutes,** or until heated through. *Serve over spaghetti squash or toasted English muffins.* Garnish with parmesan cheese. Makes 4 to 5 servings.

TUNA MICROQUETTES

- 2 (6½-ounce) cans tuna, drained
- 2 slices whole wheat bread, crumbed
- 1 egg
- 1 (10¾-ounce) can cream of mushroom soup
- 2 tablespoons lemon juice
- ⅓ cup chopped green onions with tops
- ⅛ teaspoon ground rosemary
- 1 cup frozen peas
- 2 tablespoons mayonnaise
- ½ cup sliced black olives

Combine tuna, bread crumbs, egg, ½ can soup, 1 tablespoon lemon juice, onions and rosemary in a mixing bowl. Fold in peas and divide mixture into six (6-ounce) custard cups. Arrange cups in a circle on a microwave-safe tray and cover with wax paper. **MW on High 6 to 8 minutes,** rotating tray midway through cooking. Let stand while making sauce.

In a 2-cup glass measure, combine remaining half can of soup with 1 tablespoon lemon juice, mayonnaise and olives. **MW on 70% (Medium-high) 2 minutes,** or until hot. Turn "Microquettes" out onto plate and top with sauce. *Garnish with parsley.* Makes 6 servings.

NOTE: *May also be microwaved in plastic muffin pans. Yields 8.*

VEGETABLES
RICE
DRESSINGS

Artichokes

FRESH ARTICHOKES
Thistle epistle

Prepare artichokes by slicing off stem and ½-inch of top. Snip tips off with scissors and wash artichokes under running water. Wrap each artichoke in plastic wrap, tucking ends under bottom of artichoke. Place on microwave rack. **MW on High** allowing:

1 artichoke	4 minutes
2 artichokes	7 minutes
3 artichokes	10 minutes
4 artichokes	12 minutes

Rotate midway through cooking. When finished, let stand, wrapped 5 minutes before removing the fuzzy area covering the base of each artichoke. Force top of artichoke open and use a spoon to remove the purple-tinged area. Artichokes are then ready to be stuffed or served with a sauce, such as Creamy Basil Dip, page 125.

ARTICHOKE HALVES WITH DILL BUTTER
How to serve four with two

- 2 fresh artichokes
- ¼ cup butter
- ½ teaspoon dill weed

Prepare artichokes by slicing off stem and ½-inch of top. Snip leaf tips off with scissors and wash artichokes under running water. Using a sharp knife, slice artichokes in half from top to bottom. Place four halves cut-side down in an 8x8x2-inch glass dish. Arrange artichokes so that bottom ends are toward the outside of dish and tops are in the center of dish. Cover dish tightly with plastic wrap. **MW on High 7 to 9 minutes,** rotating dish midway through cooking. Let stand at least 10 minutes before serving.

If you have small microwave-safe ramekins, divide butter into fourths among 4 ramekins. Sprinkle with dill weed and microwave on High until melted. Serve artichoke halves on plate with a ramekin of dill butter. You can also melt butter in a 1-cup measure, add dill weed and pour over artichoke halves. Makes 4 servings.

> **Asparagus**

FRESH ASPARAGUS

Simple but elegant fresh flavor

- 1 pound fresh asparagus
- 1 tablespoon lemon juice
- ¼ cup melted butter or margarine

Snap off white portion at bottom of each asparagus spear. Especially large or tough spears can be peeled at the base using a vegetable peeler. Wash thoroughly but do not dry. Arrange in an au gratin dish or glass loaf pan, with tips of asparagus to the center and stems to the outer edges. Cover with plastic wrap and **MW on High 6 to 7 minutes.** Let stand, covered, 3 minutes. Combine lemon juice and melted butter and drizzle over asparagus. Makes 4 servings.

VEGGIE TOPPER

- 3 slices bacon, diced
- ⅛ teaspoon garlic powder
- ⅓ cup dry bread crumbs or Ritz cracker crumbs

Place bacon in a 2-cup glass measure. Cover with paper towel. Stirring midway through cooking, **MW on High 3½ to 4½ minutes,** or until crisp. Remove bacon and set aside.

Discard all but 2 tablespoons bacon drippings. Combine garlic powder, crumbs and reserved bacon with bacon drippings. Mix thoroughly. Sprinkle over cooked green vegetables, such as asparagus, broccoli, Brussels sprouts or green beans. Yields ⅔ cup.

CREAMY BASIL DIP FOR ARTICHOKES

Delicious served with other veggies, too!

- 1 (3-ounce) package cream cheese
- 2 green onions, finely sliced
- 2 tablespoons parmesan cheese
- 1 teaspoon basil
- ½ cup milk
- ⅛ teaspoon worcestershire sauce

Place cream cheese in a 2-cup glass measure. **MW on 50% (Medium) 1 to 1¼ minutes.** Blend in remaining ingredients. **MW on 50% 2½ to 3 minutes,** stirring midway through cooking. Yields 1 cup. Serve warm.

ITALIAN GREEN BEANS

A change from skinny beans

- 2 (9-ounce) packages frozen Italian green beans
- 3 green onions and tops, sliced
- 2 tablespoons margarine
- ½ red bell pepper, thinly sliced
- ½ teaspoon mixed Italian seasoning

Stand packages of frozen beans in a 1½-quart round casserole. **MW on High 8 minutes.** Remove beans from boxes and place in same casserole. Add onions, margarine, red bell pepper and seasoning. Cover tightly with plastic wrap. **MW on High 3 to 4 minutes,** or to desired doneness. Makes 6 servings.

BASIL BEANS

- 2 slices bacon, diced
- ½ cup thinly-sliced onion
- 1 (9-ounce) package frozen French-style green beans
- 2 tablespoons chopped pimiento
- ¼ teaspoon salt
- ¼ teaspoon basil
- ⅛ teaspoon pepper
- ½ cup plain yogurt (optional)

Place bacon in a 1-quart casserole. Cover with glass lid. **MW on High 2 to 3 minutes.** Remove cooked bacon and place onion in drippings. Cover and **MW on High 2 minutes.**

Place frozen package of beans on a paper towel. **MW on High 5 minutes.** Add cooked beans to onion in casserole. Add reserved bacon, pimiento, salt, basil and pepper. Cover and **reheat on High** to serving temperature. Stir in yogurt, if desired. *Do not reheat after adding yogurt, because it will curdle.* Makes 4 servings.

CELERY GREEN LIMAS

Add flavor to plain limas.

- 1 (10-ounce) package frozen baby lima beans
- ⅓ cup water
- 1 cup sliced celery
- 2 tablespoons margarine
- 2 teaspoons lemon juice
- Salt and pepper to taste

Place lima beans in a 1-quart casserole and pour water over beans. Cover with plastic wrap and **MW on High 7 minutes,** redistributing beans midway through cooking. Add celery, margarine and lemon juice; re-cover. **MW on High 2½ to 3 minutes.** Let stand 3 minutes; season to taste. Makes 4 to 5 servings.

CHILI BAKED BEANS

Zippy and embarrassingly easy

- 2 (16-ounce) cans pork and beans
- 1 (15-ounce) can chili with beans
- ¼ cup molasses
- 2 teaspoons instant minced onion
- 1 teaspoon chili powder, or to taste

Drain visible liquid from cans of pork and beans. In a 2-quart casserole, combine pork and beans, chili, molasses, onion and chili powder. **MW on High 8 minutes,** or until bubbling hot. Makes 6 to 8 servings.

NOTE: May substitute ½ cup sautéed chopped fresh onion for instant minced onion. To sauté, place onion in a 1-cup glass measure. Cover with plastic wrap and **MW on High 2 to 2½ minutes.** Add to beans.

DRIED BEANS

Although dried beans take just about as long to cook in the microwave oven as they do conventionally, less electricity is used. Pot watching is also reduced, since the use of the proper microwave power level eliminates boilovers.

During the drying process, 90 percent of the moisture is removed from beans. This moisture needs to be replenished by soaking the beans before cooking. Beans can be placed in the desired cooking utensil and water added to soak overnight. Allow 6 cups of water for one pound (2 cups) of dried beans. We recommend overnight soaking for optimum quality bean cookery. However, if the chef forgot to soak the beans the night before, a Speedy Microwave Soak method can be used, see below.

BASIC INFORMATION

- One pound of dried beans equals 2 cups. Beans triple in volume when soaked and cooked, so one pound yields 6 cups.
- To soak beans, use 3 cups of water for each cup of dried beans.
- For high altitude or hard water areas, increase both soaking and cooking time to get tender beans. In addition, ¼ teaspoon of baking soda can be added to the water just before cooking to aid in softening beans.
- If beans tend to foam during cooking, add a tablespoon of vegetable oil to stop the foaming.
- Slow cooking of the beans keeps them whole and lessens the tendency for them to break apart.
- Since acidic ingredients slow the softening process of beans, wait until beans are just about tender before adding tomatoes, lemon juice, etc.
- To store: Cooked beans can be refrigerated 4 to 5 days or frozen up to 6 months. Dried beans keep indefinitely in commercial package or airtight container. Do not store in refrigerator. Dried beans are an economical source of protein. Animal proteins (meat, fish, poultry, eggs and cheese) contain all eight essential amino acids and are known as complete protein. Plant proteins lack one or more amino acids and are called incomplete. By combining complementary plant proteins, we get complete proteins such as those found in red beans and rice, tortillas and refried beans, or peanut butter and crackers.

SPEEDY MICROWAVE SOAK

Wash and sort a pound of dry beans. Place in a 4-quart simmer pot and cover beans with hot tap water. Cover with lid or plastic wrap. **MW on High for 8 to 10 minutes,** or until boiling. Let stand one hour, covered. Proceed with your favorite recipe.

→ *Dried Beans*

LIMAS 'N HAM
A favorite of Ann's dad, Ray Beddow

- 1 pound dry Lima beans
- 6 cups water
- 1 cup chopped onion
- 2 cups diced cooked ham
- Salt and pepper to taste

Wash and pick over beans. Place beans in a 4-quart simmer pot. Add water and soak overnight.

Add onion and ham, cover and **MW on High 15 to 20 minutes,** or until boiling. Stir, re-cover and **MW on 50% (Medium) one hour,** or until desired doneness. Stir every half hour, adding water if beans become too dry. Add desired seasonings. Yields 1½ generous quarts.

OLD-FASHIONED BAKED BEANS

- 1 pound Great Northern dry beans
- 6 cups water
- ½ pound bacon, cooked and crumbled
- 3 tablespoons bacon drippings
- 1 cup chopped onion
- 1 cup ketchup
- ⅔ cup molasses
- ⅓ cup packed brown sugar
- 1 teaspoon dry mustard

Wash and pick over beans. Place beans in a 4-quart simmer pot. Add water and soak overnight.

Cover and **MW on High 12 to 15 minutes,** or until boiling. Stir, re-cover and **MW on 50% (Medium) 45 minutes,** or until desired doneness. Stir every half hour, adding water if beans become too dry.

Combine bacon drippings and onion in a 2-cup glass measure. Cover and **MW on High 3½ to 4½ minutes.** Add onions, bacon, ketchup, molasses, sugar and mustard to beans. Stir and **MW on 50% 4 to 5 minutes,** or until beans are heated through. Yields 1½ generous quarts.

Dried Beans

NEW ORLEANS RED BEANS

An inspiration from our friend, Mary Beth Cyvas

- 1 pound red kidney beans
- 8 cups hot tap water
- 1 large ham bone or 1 pound ham seasoning meat
- 2 cups chopped onion
- ½ cup chopped celery
- 5 to 6 toes garlic, minced
- ¼ cup chopped parsley
- ½ cup chopped green onions
- 1 teaspoon salt
- ¼ teaspoon black pepper
- ¼ teaspoon cayenne

Wash and pick over beans. Place in a 4-quart microwave-safe pot. Add water and soak beans overnight.

Add ham bone, onions, celery, garlic, parsley and green onions. Cover and **MW on High 15 to 20 minutes,** or until boiling; then **MW on 50% (Medium) 1½ to 2 hours,** stirring every half hour. Remove bone and add seasonings. Take out 1 cup cooked beans and puree in food processor or blender. Stir into beans to thicken liquid. Yields 2½ quarts.

Broccoli

TOADSTOOLS AND FLOWERS

- 8 ounces fresh mushrooms
- 1 bunch fresh broccoli
- 2 tablespoons margarine
- ¼ cup white wine
- 1 teaspoon Italian seasoning

Clean mushrooms and cut into quarters. (Small mushrooms can be left whole). Wash broccoli and cut into flowerets. Place mushrooms and broccoli into a 2-quart rectangular glass dish.

To melt margarine, **MW on High 20 to 30 seconds.** Add wine and seasonings. Mix and pour over vegetables. Cover with plastic wrap and **MW on High 9 to 10 minutes.** Let stand 3 minutes before serving. Makes 6 to 8 servings.

BROCCOLI BUTTONS AND BOWS

- 1 large head fresh broccoli

Wash broccoli and cut tops into flowerets. Trim branches and leaves from stems and peel if desired. Weigh stems and flowerets and determine cooking time at **6 to 7 minutes per pound on High power.** Insert slicing disk into food processor. Pack stems tightly into feed tube and slice using firm pressure. Place broccoli "buttons" in the bottom of a 2-quart round casserole. Top with flowerets. Cover tightly with plastic wrap and microwave according to time you have determined. Serve with Hollandaise Sauce, below. Makes 6 servings.

FOOD PROCESSOR HOLLANDAISE SAUCE

- ½ cup butter or margarine (1 stick)
- 3 egg yolks
- 2 tablespoons fresh lemon juice
- ½ teaspoon salt
- Dash of cayenne pepper

Place butter in a 1-cup glass measure. **MW on High 1 to 1½ minutes,** or until boiling. Meanwhile, place egg yolks in food processor. Using steel blade, pulse briefly. Add lemon juice, salt and cayenne, and pulse until blended. Turn processor on and pour hot butter into feed tube in a slow, steady stream. Continue processing 20 seconds.

Remove steel blade and put processor bowl in microwave oven. **MW on 30% (Medium-low) 1 minute,** stirring with a small whisk every 15 seconds. Serve immediately. If necessary to reheat, pour Hollandaise sauce into a 2-cup glass measure set in a bowl containing 1-inch of water. **MW on 50% (Medium),** stirring frequently, until hot. Yields 1 cup.

NOTE: *If bowl of food processor has a metal handle, do not use in microwave oven. The tiny spring in the plastic handle does not harm the oven.*

Broccoli

LEMON BROCCOLI SPEARS

For the Garden State, home of our HUNTERDON COUNTY DEMOCRAT

- 1 pound fresh broccoli spears
- 1 small onion, sliced into rings
- 2 tablespoons margarine
- 1 tablespoon fresh lemon juice
- ¼ teaspoon dried tarragon
- Salt and pepper if desired

Arrange broccoli spears in a 2-quart rectangular casserole with stems toward ends of dish and flowerets in the center. Top with onion rings. Cover tightly with plastic wrap and **MW on High 8 to 10 minutes.** Do not remove cover; let stand 5 minutes. Drain well.

Place margarine in a 1-cup glass measure. **MW on High 30 to 40 seconds,** or until melted. Add lemon juice and tarragon and pour over drained broccoli. Salt and pepper as desired. Makes 6 servings.

BROCCOLI AND CARROT MEDLEY

- ¼ teaspoon garlic powder
- 2 teaspoons margarine
- 2 cups sliced carrots
- 2 tablespoons water
- 3 cups broccoli flowerets
- 1 tablespoon toasted sesame seeds

Combine garlic powder, margarine, carrots and water in a 1-quart casserole. Cover with plastic wrap and **MW on High 5 minutes.** Stir in broccoli, re-cover and **MW on High 3 to 3½ minutes.** Let stand 3 minutes. Sprinkle sesame seeds on top to serve. Makes 4 to 5 servings.

"One of my microwave class students showed me your book. It's **great!** I need one to use in teaching my microwave classes! Please send me one **RIGHT AWAY!!**"

Yvonne Murray, Kendall, Wisconsin

SMOKEHOUSE CABBAGE

Delicious with sausage or ham

- 4 slices bacon, diced
- ¼ cup finely chopped onion
- 1 pound cabbage, chopped
- ¼ cup sour cream
- ½ teaspoon poppy seed

Place diced bacon in a 1½-quart round casserole. Cover and **MW on High 2½ to 3 minutes,** or until bacon is browned. Remove bacon and reserve.

Add onion to bacon drippings and **MW on High 2½ minutes.** Add cabbage to onion, cover and **MW on High 8 minutes,** stirring midway through cooking. Blend in sour cream and poppy seed. Garnish top with reserved bacon. **MW on 70% (Medium-high) 1½ to 2 minutes,** or until heated through. Makes 5 to 6 servings.

BAVARIAN CABBAGE

A German specialty in honor of our MILWAUKEE SENTINEL

- 2 tablespoons bacon drippings or butter
- 1 medium onion, thinly sliced
- 1 (1½-pound) head red cabbage, thinly sliced
- 2 tart cooking apples, cored and chopped
- ½ cup sugar
- ½ teaspoon salt
- ¼ teaspoon pepper
- 1 bay leaf
- Pinch of ground cloves
- 1 tablespoon lemon juice
- ½ cup red wine vinegar or apple cider vinegar
- 2 cups water
- 2 tablespoons flour

Combine bacon drippings and onion in a 3-quart round casserole. Cover and **MW on High 3 minutes.** Add cabbage, apples, sugar, salt, pepper, bay leaf and cloves. Pour over lemon juice, vinegar and water and toss to combine well. Stirring every 10 minutes, cover and **MW on High 30 to 35 minutes.** Add flour and toss to absorb liquid and thicken. Makes 6 to 8 servings.

CITRUS SPICE CARROTS

Inexpensive but loaded with vitamins

1 pound carrots, peeled and cut into coins
¾ cup orange juice, divided
1½ teaspoons cornstarch
⅛ teaspoon nutmeg
1 tablespoon snipped fresh parsley

Place carrots and ¼ cup orange juice in a 1-quart casserole. Cover and **MW on High 6 to 7 minutes,** stirring once midway through cooking. Blend cornstarch and nutmeg into ½ cup orange juice. Combine spiced orange juice with carrots. Cover and **MW on High 2½ to 3 minutes,** or until sauce is thickened. Garnish with parsley. Makes 5 to 6 servings.

HONEY GINGERED CARROTS

1 pound fresh carrots
¼ cup water
¼ cup honey
2 teaspoons lemon juice
½ teaspoon ground ginger
¼ teaspoon salt
⅛ teaspoon white pepper
1 tablespoon fresh snipped parsley

Pare carrots and cut into thirds. Slice lengthwise in half, and then slice each half lengthwise. Combine carrots and water in a 1-quart casserole. Cover with plastic wrap and **MW on High 7 to 8 minutes.** Drain.

Mix honey, lemon juice, ginger, salt and pepper; toss with carrots. Sprinkle with parsley and **MW on High** until reheated. Makes 4 to 6 servings. Also great chilled!

Carrots

APRICARROTS

Hot fruited veggie for poultry or pork

- 1 pound fresh carrots
- ½ cup apricot or peach preserves
- ½ teaspoon ginger
- 1 (16-ounce) can apricot halves
- 1 tablespoon cornstarch
- ¼ cup sliced toasted almonds

Peel carrots and slice crosswise into coins. Combine carrots, preserves and ginger in a 2-quart casserole or batter bowl. Cover with plastic wrap and **MW on High 8 to 10 minutes.**

Reserve ½ cup apricot liquid. Drain apricots and cut each half into two pieces. Add to carrots. Combine reserved liquid with cornstarch. Add to carrot mixture along with almonds. Re-cover with plastic wrap and **MW on High 4 to 5 minutes,** or until liquid is thickened. Makes 4 to 6 servings.

MR. MCGREGOR'S GARDEN SPECIAL

Is this what Peter Rabbit was after?

- 4 tablespoons margarine
- ¾ cup chopped onion
- 1½ cups thinly-sliced celery
- 3 cups shredded carrots
- Salt and pepper to taste

Place margarine and onion in a 1½-quart casserole. **MW on High 3 to 3½ minutes,** stirring once. Add celery and carrots, cover and **MW on High 5 to 7 minutes,** stirring once. Add salt and pepper to taste. Makes 6 servings.

Cauliflower

INDIAN SPICED CAULIFLOWER

Authentic Hindu recipe from our friend, Monjula Chidambaram

- 1 tablespoon margarine
- 1 teaspoon coriander
- ½ teaspoon curry powder
- ¼ teaspoon ginger
- ½ cup finely chopped onion
- 3 tablespoons ketchup
- 1 head cauliflower

Combine margarine, coriander, curry and ginger in a 1-cup glass measure. **MW on High 1 to 1½ minutes,** stirring every 30 seconds. Blend in onion and ketchup.

Remove leaves and center core from cauliflower. Weigh cauliflower and place in dish large enough to hold cauliflower. Coat cauliflower with prepared spice mixture. Cover with plastic wrap and **MW on High 6 minutes per pound of cauliflower,** rotating once. Let stand 3 to 4 minutes. Makes 5 to 6 servings.

CURRY SAUCE

Delicious served over vegetables or seafood

- 2 tablespoons margarine
- 2 tablespoons flour
- ½ cup water
- ½ cup milk
- 1 teaspoon instant chicken bouillon granules
- ½ teaspoon curry powder

Place margarine in a 4-cup glass measure. **MW on High 30 seconds,** or until melted. Using a wire whisk, blend flour into margarine, and then gradually blend in remaining ingredients. **MW on High 3 to 4 minutes,** stirring with whisk midway through and at end of cooking. Yields 1 cup.

→ Celery

CRUNCHY CELERY CASSEROLE

Unusual side dish from CiCi's Aunt Dorothy

- 4 cups diagonally-sliced celery
- 1 (10½-ounce) can condensed cream of celery soup
- ½ cup milk
- ½ cup pecan pieces
- 1 cup crushed buttery crackers (such as Ritz)
- 3 tablespoons melted margarine

Place celery in an even layer in a 1½-quart casserole. Cover and **MW on High 5 to 6 minutes,** stirring once midway through cooking. Drain. Add soup, milk and pecans to cooked celery and stir until well blended. Sprinkle crushed crackers over top of celery mixture and drizzle with margarine.

NOTE: *May be made ahead and refrigerated at this point.* **MW on 70% (Medium-high) 6 to 8 minutes,** or until heated through. Makes 6 servings.

BRAISED CELERY-MUSHROOM COMBO

Low calorie and lots of fiber

- 4 cups sliced celery (½-inch diagonal) pieces
- 4 ounces fresh mushrooms, sliced
- 1 tablespoon olive oil
- 1 tablespoon lemon juice
- 1 teaspoon parsley flakes
- 1 teaspoon basil
- ⅛ teaspoon pepper
- 2 teaspoons cornstarch
- 2 teaspoons water

Place celery, mushrooms, olive oil, lemon juice, parsley flakes, basil and pepper in a 2-quart round casserole. Cover with plastic wrap and **MW on High 6 to 7 minutes.**

Combine cornstarch with water in a 1-cup glass measure. Drain liquid that collected during microwaving of celery into same 1-cup glass measure. **MW on High 1 to 1½ minutes,** or until mixture is thickened. Pour over celery mixture; toss. **MW on High 1 to 1½ minutes.** Makes 6 servings.

HOW TO MICROWAVE FRESH CORN-ON-THE-COB

Cooking corn in the husk: Remove the tough outer husks, leaving two layers of light-green husk as well as the silk intact. If microwaving more than two ears at a time, arrange wagon-wheel fashion, with the tip of the ears at the hub and the base to the outside. Microwave according to table below, turning corn upside down midway through cooking. When done, remove husk and silk for serving. This is easy to do if you stand one ear on its base. Holding the tip with one hand, use a paper towel to rub husk and silk off in a downward motion.

Cooking single ears in plastic wrap: Remove all husks and silk. Wrap individual ears in plastic wrap. (We do not recommend wrapping in waxed paper, as the corn may have a waxy taste.) For times, see table below.

Cooking corn in casserole: Remove all husks and silk. Place in rectangular 2-quart glass casserole and cover with plastic wrap. It is not necessary to add water. Microwave according to table below. Turn corn upside down and rearrange position in casserole midway through cooking.

NUMBER OF EARS	MINUTES ON HIGH POWER
1	2½ to 3
2	4 to 6
4	7 to 9
6	10 to 12

BAKED CORN ON THE COB

4 tablespoons margarine
1 teaspoon dried chives or frozen chopped chives
¼ teaspoon dry mustard
¼ teaspoon salt
⅛ teaspoon pepper
4 ears corn, husk and silk removed

Place margarine in a small bowl. **MW on 30% (Medium-low)** 40 seconds, or until softened but not melted. Stir in chives, dry mustard, salt and pepper. Divide mixture and spread over corn. Wrap each ear of corn individually in a piece of Saran Wrap or other heavy plastic wrap.

Arrange corn on a plate or dish so that one ear of corn is along each side with none in the center of dish. **MW on High 10 to 12 minutes.** Let stand 3 minutes before serving. Makes 4 servings.

COLACHE

- ½ cup chopped onion
- 2 tablespoons margarine
- 2 ears corn, cut into quarters
- 2 medium-sized zucchini (½ pound)
- 2 small tomatoes, cut in wedges

Combine onion and margarine in a 2-quart casserole. Cover and **MW on High 3 to 3½ minutes.** Stir in corn pieces, re-cover and **MW on High 3 minutes.** Stir in zucchini; re-cover and **MW on High 3 minutes.** Add tomatoes; re-cover and let stand 3 to 4 minutes. Serve a corn holder for each person to hold the hot mini ear of corn for easier eating. Makes 4 servings.

NOTE: *For ease in cutting ears of corn, make a ¼ to ½-inch cut every two inches along length of ear. Use both hands to break off each quarter beginning with the base end and work towards the tip end of the ear.*

POLKA DOT CORN CUSTARD

- ½ cup chopped green bell pepper
- 2 tablespoons margarine
- 2 tablespoons flour
- ¼ cup milk
- 2 eggs, lightly beaten
- 1 (17-ounce) can cream-style corn
- ½ cup grated sharp cheddar cheese
- 1 tablespoon chopped pimiento

Place bell pepper and margarine in a 4-cup glass measure. Cover with plastic wrap and **MW on High 2 minutes.** Add flour and stir to blend well. Blend in milk and eggs; then corn. Fold in cheese and pimiento and pour into a (9-inch) round glass dish. **MW on 70% (Medium-high) 11 to 12 minutes,** or until center is set. Let stand 5 minutes. *Dust with paprika, if desired.* Makes 4 to 6 servings.

Corn ⬅━━━━━━━━━━━━━━━━━━━━━━━━━━━━━━━━

CREOLE CORN

- 2 cups fresh corn cut from cob, or 1 (10-ounce) package frozen whole kernel corn
- 2 slices bacon
- 1 small onion, sliced thinly
- 1 (16-ounce) can stewed tomatoes, drained
- 1 bay leaf
- ⅛ teaspoon pepper
- Dash of Tabasco sauce

Put fresh corn into a covered casserole. (If using frozen corn, place unopened box on a paper towel in microwave oven.) **MW on High 4 to 5 minutes.**

Meanwhile, dice bacon and arrange in a 1-quart casserole. Cover and **MW on High 3 minutes.** Remove bacon and place onion in drippings. Cover and **MW on High 2 to 3 minutes.** Add corn, stewed tomatoes, bay leaf, pepper, Tabasco and reserved bacon. **MW on High** to desired serving temperature. Makes 6 servings.

NOTE: *Also good served cold.*

CORN AND MUSHROOMS

- 4 ounces fresh mushrooms, cleaned and sliced
- 3 green onions, sliced
- 2 tablespoons margarine
- 2 cups fresh corn cut from cob, or 1 (10-ounce) package frozen whole kernel corn
- Salt and pepper

In a 1-quart casserole, combine mushrooms, onions and margarine. **MW on High 3 minutes.** Stir in corn, cover. **MW on High 5 to 6 minutes.** Let stand 5 minutes. Season to taste. Makes 4 to 5 servings.

UNBELIEVABLY EGGPLANT

Even eggplant haters love it this way.

- 1 (1½-pound) eggplant, peeled and cut into ½-inch cubes
- ⅓ cup chopped onion
- Margarine
- 3 tablespoons flour
- 1½ cups milk
- 1 cup shredded sharp cheddar cheese (4 ounces)
- ¼ teaspoon pepper
- 1 (2¼-ounce) can sliced ripe olives, drained
- 1 cup dry herb stuffing mix (such as Pepperidge Farm)

Place eggplant and onion in a 2-quart glass batter bowl. Cover and **MW on High 8 minutes,** stirring after 4 minutes. Drain. Transfer to a 1½-quart round casserole.

Place 3 tablespoons margarine in same glass bowl. **MW on High 30 seconds,** or until melted. Blend in flour using a wire whisk. Whisk in milk. **MW on High 2 minutes;** stir with whisk. **MW on High 2 minutes,** or until mixture begins to boil; whisk. Stir in cheese until melted. Add pepper.

Stir in reserved eggplant and onions, and half the can of olives. Pour half of eggplant mixture into casserole. Sprinkle with half of stuffing. Add remaining eggplant mixture and top with remainder of stuffing. Garnish with remaining olives and dot with margarine. **MW on 70% (Medium-high) 6 minutes,** or until heated through. Makes 4 to 6 servings.

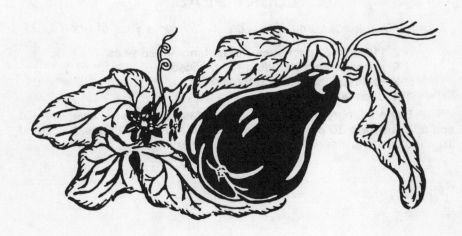

Onions

CHEESY ESCALLOPED ONIONS

1 (16-ounce) package frozen small whole onions
3 tablespoons margarine
12 Ritz crackers, finely crushed
2 tablespoons flour
½ teaspoon dry mustard
1 cup milk
¾ cup shredded cheddar cheese

Make an "X" in package of onions and place cut-side down on a paper plate. **MW on High 6½ to 7½ minutes,** re-distributing contents of package midway through cooking. Let stand in package while making sauce.

Place margarine in a 4-cup glass measure and **MW on High 45 seconds,** or until melted. Remove 1 tablespoon margarine and combine with cracker crumbs in a small bowl. Reserve.

Add flour and mustard to remaining margarine in glass measure. Stir in milk and **MW on High 3 minutes,** or until thickened, stirring midway through cooking with a wire whisk. Add cheese; stir until blended.

Drain onions and combine with cheese sauce in a 1-quart casserole. Sprinkle crumb mixture over top and **MW on 70% (Medium-high) 6 to 7 minutes.** Makes 4 servings.

LUCKY PEAS

Texans always eat some New Year's Day for a year of good luck!

2 (10-ounce) packages frozen black-eyed peas
6 green onions including tops, sliced
2 tablespoons bacon drippings
1½ cups water

Place all ingredients in a 3-quart glass casserole. Cover with lid and **MW on High 30 minutes,** stirring once after 10 minutes. Let stand 10 minutes before serving. Makes 6 to 8 servings.

➡ *Peas*

FRENCH STYLE PEAS

¼ cup thinly sliced green onions
2 tablespoons butter or margarine
1 (10-ounce) package frozen peas
½ teaspoon sugar
   Pinch of white pepper
1 cup shredded lettuce

Place onions and butter in a 1-quart casserole. **MW on High 1 minute.** Add peas, sugar and pepper. Cover and **MW on High 5 to 6 minutes.** Stir in shredded lettuce. Cover and let stand 2 minutes. Makes 4 servings.

STUFFED ONION CUPS

3 onions (about ½-pound each)
1 pound ground beef
¼ cup chopped green bell pepper
2 slices bread, crumbled
1 teaspoon worcestershire sauce
1 garlic clove, minced
½ cup chili sauce + additional to drizzle over top

Peel onions and slice off root and top edges to level. Cut onions crosswise. Scoop out centers leaving 3 layers of onion shell. Place onion cups in a 1½-quart rectangular dish. Cover with plastic wrap and **MW on High 6 to 7 minutes,** or until fork tender.

Combine ground beef, green pepper, bread, worcestershire, garlic and ½ cup chili sauce. Divide mixture among onion cups. Cover with plastic wrap and **MW on High 7 to 8 minutes,** rotating position of onions midway through cooking. Drizzle with additional chili sauce and let stand 5 minutes. Makes 4 to 6 servings.

Quick-Tip ➡ Sauté chopped vegetables such as onion and celery in a glass measuring cup covered with plastic wrap. No additional liquid is needed due to the high water content in the vegetables. If a "butter flavor" is desired, sprinkle 1 teaspoon Butter Buds over vegetables prior to sautéeing.

Potatoes ←

SPEEDY MASHED POTATOES

1 (2-pound) bag frozen Southern-style hash-brown potatoes
⅔ cup chopped onion
1¼ cups hot water
3 tablespoons margarine
¼ cup milk
Salt and pepper to taste

Combine potatoes, onions and water in glass mixer bowl. Cover with plastic wrap and **MW on High 18 to 20 minutes,** stirring midway through cooking. Do not drain. Add margarine, milk and seasonings. Whip until fluffy. Makes 6 to 8 servings.

SPEEDY GERMAN POTATO SALAD

5 slices bacon, diced
6 green onions with tops, sliced
1 pound frozen Southern-style hash brown potatotes
¼ cup wine vinegar
½ teaspoon celery salt

Place bacon in a 2-quart casserole. Cover and **MW on High 5 minutes,** or until bacon is done. Using slotted spoon, remove bacon and set aside. Add onions to bacon. Cover and **MW on High 1 minute.** Add frozen potatoes. Cover and **MW on High 10 minutes,** breaking up potatoes once or twice during cooking. Add vinegar, celery salt and reserved bacon; toss. Reheat on High to serving temperature. Makes 4 servings.

GOLDEN POTATO CASSEROLE

1 (2-pound) bag frozen Southern-style hash-brown potatoes
2 cups sour cream
2 cups shredded cheddar cheese (8 ounces)
1 bunch green onions, chopped
White pepper

Make a 1-inch slit in bag of potatoes and place bag on a paper plate. **MW on High 13 to 15 minutes,** turning bag upside down and redistributing contents every 5 minutes.

Open bag and pour potatoes into a 2-quart casserole dish. Blend in sour cream, cheese, onions and pepper to taste. **MW on 70% (Medium-high) 4 minutes.** Stir, sprinkle top of casserole with paprika and **MW on 70% 2 to 3 minutes,** or until heated through. Makes 8 servings.

KARTOFFELRÖSTI
Crusty shredded potatoes from Switzerland

- 2 pounds baking potatoes
- 4 tablespoons butter
- ¾ teaspoon salt
- 2 tablespoons water

Pierce each potato once with a fork or knife. Arrange in a circular pattern on a microwave meat or bacon rack. **MW on High 12 to 14 minutes,** rotating rack midway through cooking. Let stand until cold, or refrigerate to assemble the following day.

Peel potatoes and shred coarsely. Melt butter in a medium-size skillet on a **conventional range top**. Add shredded potatoes; sprinkle with salt and water. Using a spatula, press potatoes into a flat cake. Cover and cook over medium heat about 15 to 20 minutes, or until potatoes are golden and crusty on the bottom. Turn onto a hot serving plate, crusty side up, and serve immediately. Makes 4 servings.

BAKED POTATO SKINS

- 6 baking potatoes (about 2 pounds)
- 3 tablespoons melted butter or margarine
- Grated parmesan cheese
- Garlic powder
- Celery salt
- Mixed Italian seasoning

Wash and pierce each potato once with a fork. Arrange on a bacon rack in a circle. **MW on High 12 to 14 minutes,** rotating rack and turning each potato over midway through cooking. Let cool.

Cut each potato in two lengthwise; then cut each part in half lengthwise. Using a spoon, scoop potato out, leaving ¼-inch thickness on skin. Arrange skins on a round microwave-safe tray. Drizzle with melted butter and sprinkle with desired amount of cheese, garlic powder, salt and seasoning. Dust with paprika, if desired. Do not cover. **MW on High 12 to 14 minutes,** rotating tray and rearranging individual skins if some begin to overcook. Yields 24 potato skins.

Quick-Tip ➡ Use leftover potato scooped from skins to make Kartoffelrösti (this page, above) or Gourmet Potatoes on page 151 in our *Microwave Know-How* book, *MicroScope Savoir Faire*. (See coupon in back for ordering information.)

DILLY CHEESE POTATOES

Slice the potatoes in your food processor.

- 1 (10½-ounce) can condensed cream of onion soup
- ¾ cup milk
- 1 cup shredded cheddar cheese
- ½ teaspoon dill weed (dried)
- ¼ teaspoon pepper
- 4 cups sliced potatoes (about 1¼ pounds)
- 2 tablespoons grated parmesan cheese

In a 2-quart round casserole, combine soup, milk, cheese, dill weed and pepper. Add sliced potatoes, stirring well to coat. Cover with casserole lid or plastic wrap. **MW on High 10 minutes.** Stir to rearrange slices of potatoes. Cover and **MW on High 8 to 10 minutes.** Sprinkle with parmesan cheese and dust with additional dill weed. Let stand 5 to 10 minutes before serving. Makes 4 to 5 servings.

SWEET POTATO POTS

Make ahead in individual ramekins.

- 1 (29-ounce) can sweet potatoes, drained
- 1 (8-ounce) can crushed pineapple, drained
- 1 egg
- ¼ cup packed brown sugar
- ¼ teaspoon cinnamon
- ⅛ teaspoon nutmeg
- ½ cup miniature marshmallows
- 6 maraschino cherries

Thoroughly mash sweet potatoes by hand or in food processor. Add pineapple, egg, sugar, cinnamon and nutmeg. Blend well. Fold in marshmallows. Divide among six ramekins or place in a 1-quart casserole dish. Arrange ramekins in a circle on a microwave-safe tray. Cover with waxed paper and **MW on High 4 to 5 minutes,** rotating tray midway through cooking. Garnish with cherries. Makes 6 servings.

Sweet Potatoes

SWEET POTATOES CALYPSO

Perfect with poultry or ham

- 1 (30-ounce) can sweet potatoes, drained
- 1 (8-ounce) can crushed pineapple packed in juice
- ¼ cup orange juice
- ⅓ cup packed dark brown sugar
- 2 tablespoons margarine, melted
- 1 orange

Coarsely mash potatoes in a 2-quart round casserole. Add pineapple with juice, orange juice, sugar and margarine. Grate rind of orange into casserole and combine with other ingredients. Thinly slice orange and arrange over potato mixture. **NOTE:** *At this point, you may cover and refrigerate up to 2 days.* When ready to cook, uncover and **MW on High 8 to 10 minutes,** or until heated through. Makes 6 servings. *Recipe may be doubled.*

CANDIED-APPLE YAMS

Layered fruit and potato casserole

- 1 (30-ounce) can sweet potatoes, drained and sliced
- 2 large cooking apples
- ½ cup packed dark brown sugar
- 1 tablespoon cornstarch
- ½ teaspoon cinnamon
- 4 tablespoons margarine
- Maraschino cherries for garnish (optional)

Place half the sliced sweet potatoes in a round 2-quart casserole. Quarter, core and peel 1 apple. Make 4 slices of each quarter and place on top of potatoes. In a small bowl, combine sugar, cornstarch and cinnamon until blended very well. Sprinkle half of mixture over apples and sweet potatoes. Dot with half the margarine. Repeat layers with remaining sweet potatoes, apple, sugar mixture and margarine. Cover and **MW on High 9 to 10 minutes.** Add cherries, if desired, and let stand 5 minutes before serving. Makes 6 servings. *Recipe may be doubled.*

SPINACH CORIANDER

In the Greek manner

- 1 (10-ounce) package frozen chopped spinach
- ¼ teaspoon turmeric
- ¼ teaspoon ground coriander
- ½ cup creamy cottage cheese
- ¼ cup dry herb stuffing mix

Stand package of spinach in a 1-quart casserole. **MW on High 5 to 6 minutes.** Press all liquid from package using the back of a wooden spoon. Discard liquid and place spinach in same casserole. Add turmeric, coriander and cottage cheese and combine well. Top with stuffing mix. **MW on 70% (Medium-high) 5 minutes,** or until heated through. Makes 4 servings.

CHEESY SPINACH

Doubles as a meatless entree over toast

- 2 (10-ounce) packages frozen chopped spinach
- 2 tablespoons margarine
- 2 tablespoons flour
- 1 cup milk
- 1 teaspoon worcestershire sauce
- 1½ cups shredded sharp cheddar cheese (6 ounces)
- 4 hard-cooked eggs, sliced or chopped
- Paprika

Place unwrapped packages of spinach in a 1½-quart round casserole. Cover and **MW on High 9 to 10 minutes,** redistributing blocks of spinach midway through cooking. Do not discard liquid.

Place margarine in a 4-cup glass measure. **MW on High 30 seconds.** Whisk in flour, then milk and worcestershire sauce. Whisking midway through cooking, **MW on High 3 to 4 minutes,** or until thickened. Stir in cheese until melted.

Pour sauce into spinach and liquid; blend. **MW on 70% (Medium-high) 3 to 4 minutes,** or until heated through. *Serve over toast points or English muffins.* Garnish top of individual servings with egg; dust with paprika. Makes 4 servings.

CRUNCHY ACORN SQUASH

Three-ingredient easy!

- 2 acorn squash, about 1-pound each
- Margarine
- 1/3 cup granola

Cut squash in half lengthwise. Remove seeds and membranes. Place cut-side down in a 10-inch square casserole. Cover with plastic wrap and **MW on High 10 minutes.**

Turn squash cut-side up and place a generous teaspoonful of margarine in each half. **MW on High 1 minute,** or until margarine is melted. Divide granola among squash halves and stir into melted margarine. Cover with plastic wrap and **MW on High 2 minutes,** or until heated through. Makes 4 servings.

CRANAPPLE ACORN SQUASH

Fantastic fruit filling

- 2 acorn squash, about 1½ pounds total
- 2 medium apples (McIntosh, Winesap or Granny Smith)
- 2 tablespoons packed dark brown sugar
- ½ cup whole cranberry sauce
- ½ teaspoon cinnamon

To make squash easier to cut, **MW on High 2 minutes.** Cut in half end to end; remove seeds and strings. Place in an 8-10" square baking dish.

Quarter, core and peel apples. Dice into ½-inch pieces. Combine apples, sugar, cranberry sauce and cinnamon. Divide mixture into centers of the four squash halves. Cover dish with plastic wrap. **MW on High 10 to 12 minutes,** or until squash can be pierced easily with a fork. Let stand 5 minutes before serving. Makes 4 servings.

Squash

ZUCCHINI BOATS

Yummy dugout canoes

- 3 medium zucchini (about 1 pound)
- 4 ounces fresh mushrooms, cleaned and chopped
- ¼ cup finely chopped onion
- 1 tablespoon margarine
- ⅓ cup Italian-flavored bread crumbs
- ⅓ cup shredded mozzarella cheese
- Paprika

Cut off ends of zucchini, pierce skin once and place in a 1½-quart rectangular dish. Cover with waxed paper and **MW on High 2½ to 3½ minutes,** redistributing zucchini once. Let stand 3 to 4 minutes.

Combine mushrooms, onion and margarine in a 2-cup glass measure. **MW on High 2 to 2½ minutes.**

Cut zucchini in half lengthwise, scoop out pulp and add pulp to mushroom mixture. Blend in bread crumbs and cheese. Place this mixture back into zucchini halves. Sprinkle with paprika and **MW on High 3 to 4 minutes,** or until heated through. Makes 4 servings.

STOPLIGHT SPECIAL

Red, yellow and green

- ½ cup thinly sliced green onions
- 2 tablespoons margarine
- ½ teaspoon basil
- ½ pound zucchini
- ½ pound yellow crookneck squash
- 1 large firm tomato, thinly sliced

Combine onion, margarine and basil in a 1-cup glass measure. **MW on High 1 to 1½ minutes,** or until margarine is melted.

Slice squash ¼-inch thick. Place zucchini in bottom of a 1-quart casserole. Drizzle one-third of onion mixture over top. Repeat layers, using yellow squash and then tomato slices. Top with remaining onion mixture. Cover and **MW on High 6 to 8 minutes,** rotating dish once. Let stand 5 minutes before serving. Makes 4 to 5 servings.

ZUCCHINI TOMATOES

Ratatouille without eggplant

- ½ cup sliced onion
- 4 tablespoons olive oil
- 1 tablespoon chopped parsley
- 1 pound zucchini squash (3 to 4 small)
- 2 fresh tomatoes (about ½ pound)
- ¼ teaspoon basil
- Dry bread crumbs
- Grated Parmesan cheese
- Paprika

Combine onion and olive oil in a 1-cup glass measure. Cover with plastic wrap and **MW on High 2 to 2½ minutes.** Add parsley, stir and set aside.

Slice squash into ¼-inch slices. Place half of slices on bottom of a 1-quart round casserole. Cover with half of onion mixture. Slice one tomato and arrange slices over onion mixture. Sprinkle with half of basil. Repeat layers of squash, onion mixture and sliced tomato. Sprinkle top tomato layer with basil. Top with dry bread crumbs, cheese and paprika. **MW on High power 7 to 8 minutes.** Let stand 3 minutes before serving. Makes 6 servings.

VEGGIE FONDUE

Children love this!

- Assorted raw vegetables cut into pieces (such as carrots, cauliflower, broccoli, celery, summer squash and whole cherry tomatoes)
- 1 (11-ounce) can condensed cheddar cheese soup
- ½ of a soup can of milk
- 1 cup shredded Swiss cheese (4 ounces)
- ½ teaspoon worcestershire sauce
- ⅛ teaspoon garlic powder

Wash vegetables and refrigerate until needed. Combine soup, milk, cheese, worcestershire sauce and garlic powder in a 1-quart casserole. **MW on 70% (Medium-high) 5 to 6 minutes,** stirring once midway through cooking. Use fondue forks for dipping vegetables into cheese fondue. Yields 2 cups.

Vegetables

VEGGIE POTPOURRI

In honor of THE EAGLE, hometown newspaper of the Texas A&M Aggies

- ½ medium zucchini, sliced ¼-inch thick
- ¼ green bell pepper, sliced
- 6 fresh mushrooms, sliced
- ¼ yellow onion, sliced
- 2 ribs celery, sliced
- 1 tablespoon margarine
- ½ teaspoon Italian seasoning
- 1 tomato, cut into wedges

In a 1-quart casserole, place zucchini, bell pepper, mushrooms, onion, celery, margarine and seasoning. Cover and **MW on High 4 to 5 minutes.** Add tomato, and re-cover. **MW on High 1 minute.** Let stand 3 minutes. Makes 4 servings.

FRESH VEGGIE KABOBS

Also good for outdoor grilling

- 1 small head cauliflower
- 2 medium zucchini
- 1 large green bell pepper
- 4 ounces fresh mushrooms
- ½ pint cherry tomatoes
- ½ cup Italian dressing

Prepare vegetables by cutting cauliflower into bite-size flowerets. Cut zucchini into 1-inch slices and bell pepper into 1-inch pieces. Combine all vegetables in a large mixing bowl and pour dressing over top. Marinate at least one-half hour.

Reserve cherry tomatoes. Thread remaining vegetables onto (8 to 10-inch) bamboo skewers, leaving space at each end for tomatoes to be added later. Place skewers in a 2-quart rectangular dish. Cover with plastic wrap and **MW on High, allowing 30 to 45 seconds per skewer of vegetables.** Place one tomato on each end of skewers, re-cover and **MW on High 1 to 1½ minutes,** or until tomatoes are heated. Makes 5 to 6 servings.

REGULAR WHITE RICE

- 2 cups hot tap water
- 1 teaspoon salt
- 1 tablespoon margarine
- 1 cup uncooked rice

In a 2-quart casserole combine water, salt and margarine. Cover and **MW on High 4 minutes,** or until boiling. Stir in rice. Cover and **MW on High 3 minutes,** then on 30% (Medium-low) 12 to 14 minutes. Fluff rice with a fork, re-cover and let stand 5 minutes. Yields 3 to 3½ cups.

INSTANT RICE

Refer to package for desired number of servings. Place recommended amount of water, margarine and salt into a casserole. Cover and **MW on High** until water comes to a full rolling boil. Stir in rice, re-cover and let stand 5 to 8 minutes for water to be absorbed. Fluff with fork. Before serving, reheat on High if necessary.

SPANISH RICE

Add chilies if you like it hot.

- 1 (16-ounce) can stewed tomatoes
- ½ cup water
- 1 tablespoon margarine
- ½ teaspoon garlic powder
- ½ teaspoon cumin (or to taste)
- 1 cup regular uncooked rice
- ¾ cup shredded cheddar cheese (optional)

Whirl tomatoes in blender or food processor. Combine with water, margarine, garlic powder and cumin in a 2-quart casserole. Cover and **MW on High 4 minutes,** or until boiling. Stir in rice. Cover and **MW on High 3 minutes,** then on 30% (Medium-low) 14 to 16 minutes. Fluff rice with a fork. Sprinkle cheese over top; re-cover and **MW on 70% (Medium-high) 1½ minutes,** or until cheese is melted. Let stand 5 minutes before serving. Makes 4 to 5 servings.

ORANGE RICE

A lovely bed for curries

- 1 cup orange juice
- 1 cup water
- 1 tablespoon margarine
- 1 cup uncooked white rice
- ⅛ teaspoon cinnamon
- ⅛ teaspoon allspice
- ⅛ teaspoon nutmeg
- ¼ cup raisins

In a 2-quart casserole, combine orange juice, water and margarine. Cover and **MW on High 4 minutes,** or until boiling. Stir in remaining ingredients. Cover and **MW on High 3 minutes, then on 30% (Medium-low) 12 to 14 minutes.** Fluff rice with a fork, re-cover and let stand 5 minutes. Makes 5 to 6 servings.

FRIED RICE

A Chinese restaurant at home

- ¼ pound bacon
- 2 tablespoons bacon drippings
- 3 green onions, thinly sliced
- ¼ cup chopped green bell pepper
- 3½ cups cooked rice, chilled
- 1 egg
- 1 tablespoon water
- 2 teaspoons soy sauce

Place bacon strips on a bacon or meat rack. Cover with paper towels and **MW on High 3½ to 4½ minutes,** or until bacon is desired doneness. Crumble bacon and reserve.

Combine bacon drippings, green onions and green pepper in a 1½-quart casserole. Cover and **MW on High 2½ to 3 minutes.** Blend in rice.

Place egg and water in a 1-cup glass measure. Whisk with a fork. **MW on High 45 to 60 seconds,** stirring midway through cooking. Use fork to break apart cooked scrambled egg. Add to rice mixture. Sprinkle soy sauce over top and add reserved bacon. Blend ingredients. Cover and **MW on High 4 to 5 minutes,** or until heated through. Makes 4 to 5 servings.

➤ *Stuffings*

DIRTY RICE DRESSING

A Louisiana specialty

- 1 pound bulk sausage, regular or hot flavor
- 1 medium green bell pepper, chopped
- 3 cups water
- 1½ cups raw long grain rice
- ½ teaspoon instant minced garlic
- ¼ teaspoon cayenne pepper (optional)
- 6 green onions, thinly sliced
- 1 cup beef bouillon or stock
- 4 tablespoons margarine
- Chopped cooked giblets (optional)
- Salt and pepper to taste

Crumble sausage into a hard-plastic colander set in a 3-quart casserole. Sprinkle bell pepper over top. **MW on High 3 minutes.** Stir well to break up sausage. **MW on High 3 minutes,** or until sausage is no longer pink; stir. Set colander aside and discard fat from casserole.

Put water in same casserole. Cover and **MW on High 5 to 6 minutes,** or until water is boiling. Stir in rice, garlic and cayenne. Cover and **MW on High 3 minutes; then on 30% (Medium-low) 13 to 15 minutes,** or until water is absorbed.

Fluff rice with fork. Add reserved sausage, bell pepper, green onions, bouillon and margarine. Mix well and add giblets, salt and pepper as desired. Use as a stuffing or reheat as a side dish. Yields 2 quarts.

Dear Microscope,
Is there something wrong with my microwave oven if it gets steam on the window?

Dear Microcook

It is normal for moisture to condense on the inside of the window if you are microwaving a food having a high moisture content. This is because there is no dry heat in the microwave oven to evaporate excess moisture as happens in conventional ovens.

If you let food remain in the microwave oven after cooking, the window is certain to steam up because the vent fan which normally circulates air will be off. This can be beneficial in oven cleaning, because the steam loosens spatters.

Stuffings

OYSTER DRESSING

In honor of our Norfolk VIRGINIAN-PILOT

- ½ cup margarine
- 1 cup chopped celery
- 1½ cups chopped onion
- ¼ cup chopped fresh parsley
- 1 clove garlic, minced
- 2 (10-ounce) jars fresh oysters, rinsed, drained and chopped
- 6 cups herb-seasoned dry stuffing
- 2 teaspoons seasoned salt
- ½ teaspoon pepper
- ½ teaspoon crushed rosemary
- 2 cups chicken or turkey stock

Place margarine in a 3-quart glass dish. **MW on High 45 seconds.** Add celery and onions; stir and arrange in a doughnut shape. Cover and **MW on High 3 minutes.** Stir in parsley, garlic, oysters, stuffing, salt, pepper, rosemary and stock. Cover and **MW on High 3 minutes.** Use as stuffing or reheat as a side dish. Yields 2 quarts.

APPLE NUT DRESSING

Delectable dressing that's good for you

- ¼ cup minced onion
- ½ cup margarine
- 2 quarts cubed day-old bread (white, whole wheat, or mixed)
- 1 teaspoon poultry seasoning
- 1 teaspoon salt
- 1 pound cooking apples
- ½ cup white raisins
- ½ cup toasted sliced almonds
- 1 to 1½ cups poultry stock

Place onion and margarine in a 2-cup glass measure. Cover and **MW on High 2 to 3 minutes.** Drizzle over cubed bread in a 3-quart casserole. Sprinkle with poultry seasoning and salt; toss well.

Quarter and core apples, but do not peel. Chop in food processor using steel blade; or coarsely chop with a knife. Add apples, raisins, and almonds to bread mixture. Moisten with stock in desired amount. Use as a stuffing, or **MW on High 8 to 9 minutes** as a side dish. Yields 2 quarts.

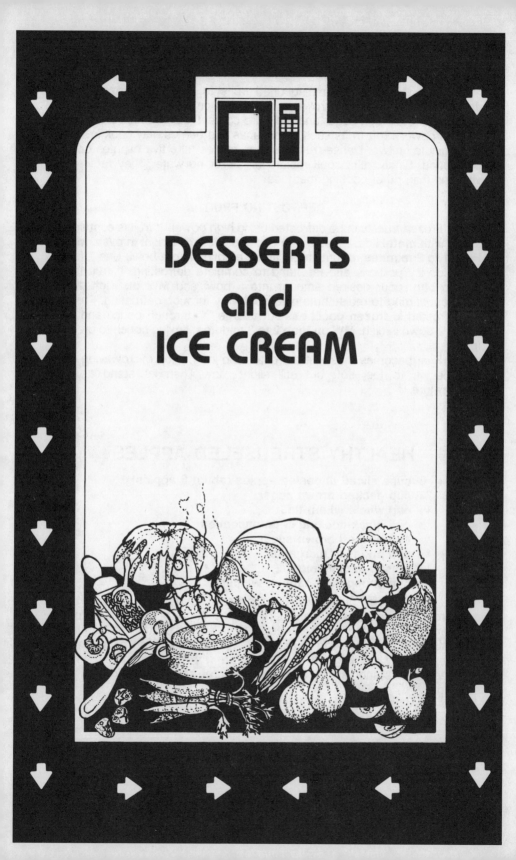

FRUIT

Everything from plumping dried fruits to defrosting frozen ones is done so easily in the microwave oven. Cooking fruits is especially fast because of their high sugar and moisture content. All fruits are microwaved on high power. Soft fruits such as bananas and berries may cook in as little as a minute or two. Dense fruits such as apples take five minutes or more per pound. Since fruits cook rapidly in little or no water, they retain more vitamins than other cooking methods.

DEFROSTING FRUIT

Frozen fruits can be defrosted using high power. If fruit is contained in a **box with metal lids,** remove one end and set box upright in oven. **MW on High 1 to 2 minutes,** or until fruit can be emptied into a bowl. Use a fork to break up icy portions and let stand to complete defrosting. If fruit is in a **plastic bag,** pour desired amount into a bowl and **MW on High 1 to 4 minutes,** stirring to redistribute pieces midway through defrosting. Fruit can be defrosted in **frozen pouches.** Cut a large "X" through pouch and place cut-side down in dish. **MW on High 2 to 3 minutes,** flexing pouch to break up fruit.

Fruit becomes mushy if defrosted too long. Stop microwaving when fruit begins to feel soft but still slightly icy. Then let stand at room temperature.

HEALTHY STREUSELED APPLES

- 6 cups sliced unpeeled apples (about 6 apples)
- ¾ cup packed brown sugar
- ⅓ cup whole wheat flour
- ⅓ cup quick-cooking oats, uncooked
- ⅓ cup packed brown sugar
- ¼ cup butter or margarine
- ½ teaspoon cinnamon
- ¼ cup chopped pecans or walnuts

Toss sliced apples with ¾ cup brown sugar in a 2-quart round casserole. To make streusel topping, use food processor or pastry blender. Combine flour, oats, ⅓ cup brown sugar, butter, and cinnamon and mix until crumbly. Add pecans and sprinkle over top of apples. Do not cover. **MW on High 8 to 10 minutes,** rotating midway through cooking. Let stand a few minutes before serving. Makes 6 to 8 servings.

> *Apples*

RED CANDIED APPLES

- ⅓ cup red hot cinnamon candies
- 1½ cups sugar
- ⅜ cup light corn syrup
- ½ cup water
- 6 medium-large red apples, room temperature

Combine candies, sugar, corn syrup and water in a 2-quart glass batter bowl. **MW on High 3 minutes.** Stir well. Without stopping the oven or stirring the mixture, **MW on High 12 minutes.** Test with candy thermometer to see if mixture has reached the hard-cracked stage (300° F.). If more cooking is needed, **MW on High 1 minute at a time,** testing after each minute, until hard-crack stage is reached.

While candy is cooking, prepare apples. Remove stems, wash apples and dry very well. Place stem-side down on a sheet of aluminum foil. Insert a wooden stick into end. When candy mixture has finished cooking, tilt bowl and twirl each apple into hot mixture. Work quickly so that mixture will not harden in bowl. Set apples on foil to cool. Makes 6 servings.

NOTE: *Set glass bowl on a heatproof surface to cool completely before attempting to wash. Adding water to hot bowl will cause it to crack, since the candy mixture has heated it to 300°.*

CANDIED APPLE RINGS

Great with poultry and pork

- ¼ cup water
- ¼ cup light corn syrup
- ½ cup red hot cinnamon candies
- 1½ pounds apple, cored, peeled and sliced in ½-inch rings

In a 2-cup glass measure, combine water, corn syrup and candies. **MW on High 4 minutes,** stirring midway through cooking. Skim uncooked candy from syrup.

Place prepared apples in an 8x8x2-inch dish. Pour cinnamon syrup over apples, cover with waxed paper and **MW on High 9 to 10 minutes.** Midway through cooking, tilt dish to baste apples with syrup, re-cover and continue cooking. Baste apples at end of cooking and let stand 5 minutes before serving. Makes 4 to 5 servings.

RUM BANANAS

Foster didn't have a microwave oven

- ½ cup packed light brown sugar
- ¼ cup rum
- ¼ cup butter or margarine
- 2 large firm ripe bananas

Combine brown sugar and rum in a 9-inch round glass dish. Add butter. **MW on High 2 minutes** and stir to combine ingredients. **MW on High 2 to 3 minutes,** or until sugar is melted. Cut bananas in half crosswise and lengthwise so that there are 8 pieces. Add to sugar mixture, coating each piece. **MW on High 1 minute.** Serve over ice cream. Makes 4 servings.

FROZEN CHOCOLATE BANANAS

When bananas are on sale, freeze some like this.

- 2 (1-ounce) squares semi-sweet chocolate
- 2 tablespoons margarine
- ¼ cup light corn syrup
- ¼ cup sugar
- 2 tablespoons milk
- Bananas
- Chopped nuts, graham cracker crumbs, or wheat germ

Combine chocolate and margarine in a 2-cup glass measure. **MW on 70% (Medium-high) 2 minutes.** Stir. Add corn syrup, sugar and milk; stir. **MW on 70% 2 minutes.** Refrigerate until completely cooled. Yields ¾ cup sauce.

Cut bananas in half, or thirds if quite large. Place a popsickle stick in one end for a handle. Hold over sauce and spoon sauce on all sides. Roll chocolate-covered bananas in dish of chopped nuts, graham cracker crumbs, or wheat germ. Place on waxed paper-lined cookie sheets. Freeze.

→ *Fruit*

BAKED GRAPEFRUIT

Please use our Texas Ruby Reds!

Grapefruit halves
2 teaspoons dark brown sugar for each grapefruit half
Cinnamon

Remove seeds and cut around sections of fruit. Sprinkle with brown sugar. **MW on High** according to chart below. Sprinkle with cinnamon before serving.

AMOUNT	TIME
1 grapefruit half	1 to 2 minutes on High power
2 grapefruit halves	2 to 3 minutes on High power
4 grapefruit halves	3½ to 4½ minutes on High power

NECTARINE COBBLER

5 cups sliced fresh nectarines (about 2 pounds)
1 (4-serving size) package non-instant butterscotch pudding mix
4 tablespoons margarine
½ cup flour
⅔ cup quick oats
½ cup packed dark brown sugar
⅓ cup chopped pecans
1 teaspoon cinnamon

Place nectarines in a 9" round glass cake dish. Sprinkle with 2 tablespoons pudding mix.

To melt margarine, place in a 4-cup glass measure and **MW on High 30 to 40 seconds.** Add remaining pudding and all other ingredients. Mix until crumbly. Distribute over nectarines. **MW on High 12 to 13 minutes,** or until bubbly, rotating dish midway through cooking. Let stand 10 minutes. Makes 6 servings.

NOTE: *Can also be made with other fruit, such as apples, pears or peaches.*

Fruit

BAKED PEARS WITH ALMONDS

Simple but elegant

- 4 firm ripe pears (about 1½ pounds)
- ¼ cup water
- ½ cup packed light brown sugar
- 2 tablespoons margarine
- ½ teaspoon cinnamon
- ¼ cup sliced toasted almonds

Cut pears in half; core and peel. Place cut-side down in an 8-inch square glass dish, arranging large ends of pears to the outside of dish. Measure water into a 1-cup glass measure. Add sugar, margarine and cinnamon. **MW on High 1 minute.** Pour mixture over pears. Cover with plastic wrap and **MW on High 6 minutes.** Let stand 2 minutes, or until ready to serve. Place 1 or 2 pear halves in each sherbet dish, spoon syrup over them, and sprinkle with almonds. Makes 4 to 6 servings.

HOT CURRIED FRUIT

For a buffet, serve in a chafing dish.

- 1 (20-ounce) can pineapple chunks, drained
- 1 (16-ounce) can pear halves or pieces, drained
- 1 (16-ounce) can sliced cling peaches, drained
- ½ cup maraschino cherries, drained
- 2 tablespoons margarine
- ⅓ cup packed light brown sugar
- 1 teaspoon curry powder

Arrange fruits in a 2-quart rectangular glass dish. Combine margarine, brown sugar and curry powder in a 1-cup glass measure. **MW on High 1 minute;** stir. Pour over fruit. Cover with plastic wrap and **MW on High 5 minutes.** Serve hot as an accompaniment to ham, pork or poultry. Makes 6 to 8 servings.

Desserts

STRAWBERRIES FLORIDIAN

In honor of our TAMPA TRIBUNE

- 1 quart fresh strawberries
- 3 tablespoons margarine
- ¼ teaspoon orange extract
- 36 Ritz crackers, crushed (1¼ cups)
- ¼ cup packed light brown sugar
- 2 cups whipped topping
- 2 tablespoons grand marnier, optional

Wash and hull berries. Place berries, stem end down, in a single layer on bottom of a 9-inch round microwave layer cake pan.

Place margarine in a 4-cup glass measure and **MW on High 45 seconds,** or until melted. Stir in extract. Blend in cracker crumbs and sugar; mix well. Cover berries with crumb mixture and **MW on High 4 to 5 minutes,** rotating dish midway through cooking. Serve warm with topping. (If desired, grand marnier can be blended in whipped topping.) Makes 5 to 6 servings.

CHOCOLATE PEPPERMINT FONDUE

A bail-out dessert when you have no time.

- 1 (6-ounce) box thin chocolate-covered mints
- 1 (6-ounce) package semi-sweet chocolate pieces
- ⅓ cup light cream

Combine ingredients in a 1-quart round casserole. **MW on 50% (Medium) 3 to 3½ minutes.** Stir. Continue to **MW on 50% 1 to 1¼ minutes.** Serve with assorted bite-size pieces of fruit or Angel food or pound cake cubes. Yields 1⅓ cups. **To double recipe:** *Use a 1½-quart round casserole and microwave ingredients 5 to 5½ minutes before stirring. Proceed as directed. Yields 2⅔ cups.*

"Last spring I received a copy of your book as a gift from a good friend. The book is all she said it was—and more! Until then, I had used my microwave mostly for reheating. It is difficult to pick out what I like best about your book, because I really like everything about it!

Marjorie Thomas, Chicago Heights, Illinois

Sugarless Desserts

PEACH BETTY

2 (16-ounce) cans sliced peaches, packed in juice
1 tablespoon cornstarch
¾ cup quick oats, uncooked
¼ cup flour
½ teaspoon cinnamon
3 packets sugar substitute to equal 2 tablespoons sugar
3 tablespoons margarine

Drain peaches, reserving juice. Measure ¾ cup juice. Using a whisk, combine juice and cornstarch in a 1-quart round casserole. Whisking midway through cooking, **MW on High 2½ to 3 minutes,** or until thickened. Stir in peaches.

Combine oats, flour, cinnamon and sugar substitute in a mixing bowl. Cut in margarine until mixture resembles the size of small peas. Distribute crumb mixture over top of peaches. **MW on High 4 to 5 minutes,** rotating dish once. Makes 5 to 6 servings.

STRAWBERRY FLUFF

1 cup water
1 (4-serving) envelope strawberry D-ZERTA
½ (20-ounce) bag frozen strawberries (no added sugar)
1 (16-ounce) carton plain yogurt
2 egg whites
3 packets sugar substitute to equal 2 tablespoons sugar

Place water in a 4-cup glass measure and **MW on High 2½ to 3 minutes,** or until a full rolling boil. Stir gelatin into water until dissolved. Refrigerate until thickened but not set.

Place strawberries in a medium-size bowl and **MW on 30% (Medium-low) 1 minute,** or until thawed just enough to slice. Slice strawberries and add yogurt; blend.

Beat egg whites until soft peaks form. Add sugar substitute and beat until dissolved. Whip gelatin and combine with yogurt and fruit. Fold beaten egg whites into fruit mixture until blended. Spoon into individual serving dishes or parfait glasses. Makes 8 servings.

→ Sugarless Desserts

SUGARLESS CHEESECAKE

For diabetics, dieters, and everybody else!

⅓ cup margarine
1 cup (unsalted tops) saltine cracker crumbs (28 crackers)
1½ (8-ounce) packages neufchatel cheese
6 packets sugar substitute to equal 4 tablespoons sugar
½ teaspoon almond flavoring
¼ teaspoon lemon flavoring
2 eggs
½ cup low-fat cottage cheese
½ cup plain yogurt
2 packets sugar substitute to equal 4 teaspoons sugar
Nutmeg

Crust: Place margarine in a 9-inch glass pie plate. **MW on High 1 minute.** Add cracker crumbs; mix well and press firmly in bottom and on sides of pie plate. **MW on High 2½ to 3 minutes,** rotating dish midway through cooking.

Filling: Place cheese in a 4-cup glass measure. **MW on 50% (Medium) 2 minutes,** or until softened. Blend in sugar substitute, flavorings and eggs. Beat until smooth. Pour into crust. Rotating midway through cooking, **MW on 70% (Medium-high) 4 minutes,** or until just set.

Topping: Place cottage cheese in blender or food processor and process until smooth. Add yogurt and sugar substitute; blend until smooth. Pour over filling. Dust top with nutmeg. Refrigerate several hours. Makes 8 servings.

NOTE: *Cheesecake may be topped with sliced fresh fruit such as strawberries and kiwi.*

MICROWAVE SPRINGFORM PAN

To make a microwave springform pan, cut a 24-inch piece of heavy plastic wrap (such as Saran Wrap). Fold in half lengthwise and drape across sides and bottom of a 9x2½-inch round plastic layer cake pan. Cut a circle of corrugated cardboard to fit bottom of pan and place on top of plastic wrap "sling" so that baked cheesecake can easily be lifted from pan. Spray cardboard with Pam.

MARBLE CHEESECAKE

- 2 tablespoons margarine
- Chocolate sandwich cookies (Oreos) crushed to yield 1¼ cups
- 3 (8-ounce) packages cream cheese
- ¾ cup sugar
- 3 eggs
- ¾ cup sour cream
- 3 (1-ounce) squares unsweetened chocolate
- ½ cup sugar

Place margarine in a 2-cup glass measure. **MW on High 45 seconds,** or until melted. Add cookie crumbs; blend well. Press crumb mixture onto cardboard bottom in microwave springform pan (see directions above). **MW on High 1 minute,** rotating pan midway through cooking.

Place unwrapped cream cheese in glass mixer bowl. **MW on 50% (Medium) 3 to 3½ minutes,** or until softened. Beat well with electric mixer. Add ¾ cup sugar; beat well. Add eggs; beat until smooth. Add sour cream; beat until smooth.

Place unwrapped chocolate squares in a 4-cup glass measure. **MW on 50% 3 to 3½ minutes,** stirring midway through cooking. Blend in ½ cup sugar. Remove 1 cup of cream cheese mixture and blend into chocolate mixture. Pour chocolate batter on top of light batter. To create marble effect, dip a rubber spatula into mix and draw a few swirls. Transfer mixture into prepared pan. **MW on 70% (Medium-high) 11 to 13 minutes,** rotating pan every 2 minutes. Let stand until cool. Refrigerate. To remove cheesecake, lift from pan using plastic wrap sling. Makes 12 servings.

Quick-Tip ➡ Cheesecake is done when there is a slight jiggle in the center. Upon cooling, the filling will set. Cheesecakes that are microwaved too long will not be as creamy and smooth.

→ *Cheesecake*

CHIMPANZEE CHEESECAKE

Your friends will go ape over this!

- 4 tablespoons margarine
- 1¼ cups graham cracker crumbs
- 2 tablespoons sugar
- 2 (8-ounce) packages cream cheese
- ¾ cup sugar
- 3 eggs
- 2 teaspoons lemon juice
- 1 cup mashed bananas (approximately 3 medium)

Place margarine in a 10-inch deep dish glass pie plate. **MW on High 45 seconds,** or until melted. Add crumbs and sugar; blend well. Press crumb mixture onto bottom and sides of pie plate. **MW on High 2 minutes,** rotating plate midway through cooking. Set aside.

Place unwrapped cream cheese in glass mixer bowl. **MW on 50% (Medium) 2½ to 3 minutes,** or until softened. Beat well with electric mixer. Add sugar; beat well. Add eggs and lemon juice; beat until smooth. Blend in bananas. Pour mixture into prepared crust. **MW on 70% (Medium-high) 10 to 11 minutes,** rotating plate every 2 minutes. Let stand until cool. Refrigerate. *If desired, garnish each serving with sliced bananas.* Makes 8 servings.

DAIQUIRI CHEESECAKE

- 3 tablespoons margarine
- 1 cup graham cracker crumbs
- 2 tablespoons sugar
- 2 (8-ounce) packages cream cheese
- 1 cup sugar
- 2 (⅝-ounce) packets instant Daiquiri mix
- 3 eggs
- ⅓ cup rum
- 2 drops green food coloring

Place margarine in a 9-inch round glass cake dish. **MW on High 35 seconds,** or until melted. Add crumbs and sugar; blend well. Press crumb mixture in bottom only of dish. **MW on High 2 minutes,** rotating dish midway through cooking.

Place unwrapped cream cheese in glass mixer bowl. **MW on 50% (Medium) 2½ to 3 minutes,** or until softened. Beat well with electric mixer. Add sugar and Daiquiri mix; beat well. Add eggs, rum and food coloring; beat until smooth. Pour mixture into prepared crust. **MW on 70% (Medium-high) 7½ to 9 minutes,** rotating dish every 2 minutes. Let stand until cool. Refrigerate. *If desired, garnish each serving with a rosette of whipped cream and a twisted lime slice.* Makes 8 servings.

CICI'S CHOCOLATE AMARETTO CHEESECAKE

Sinfully rich – heavenly easy

- 4 tablespoons margarine
- 1¼ cups vanilla wafer cookie crumbs
- 2 tablespoons amaretto liqueur
- 8 ounces semisweet baking chocolate
- 3 tablespoons milk
- 2 (8-ounce) packages cream cheese
- 1⅓ cups sugar
- 3 eggs
- 1 cup sour cream
- ½ teaspoon cinnamon
- ½ teaspoon almond extract

Place margarine in a 9-inch round high-sided layer-cake pan which is microwave safe. **MW on High 30 to 45 seconds,** or until melted. Stir in cookie crumbs and pat mixture on bottom only. **MW on High 2 minutes.** Sprinkle amaretto liqueur over crust. Set aside.

Unwrap squares of chocolate and place in large glass mixer bowl along with milk. **MW on 50% (Medium) 4 minutes;** stir. Add cream cheese and **MW on 50% 4 minutes,** or until cheese is softened. Place bowl on mixer stand and beat mixture on medium high. Add sugar; beat well. Add eggs, sour cream, cinnamon and extract; blend. Pour into prepared crust.

MW on 70% (Medium-high) 12 to 14 minutes, rotating once midway through cooking. Center should jiggle slightly when set. Cheesecake will firm considerably after refrigerating. Chill two hours or more before serving. Makes 12 to 16 servings.

Serving Suggestion: *Serve topped with whipped cream flavored with 1 to 2 tablespoons amaretto liqueur. Garnish with sliced toasted almonds.*

➤ *Pudding*

CHRIS' BANANA PUDDING

The youngest Williamson's favorite

1 (4-serving size) non-instant vanilla pudding mix
2 cups milk
2 medium bananas
Vanilla wafer cookies
Whipped topping

Empty pudding mix into a 4-cup glass measure. Using a wire whisk, stir in milk a little at a time until blended. **MW on High 3 minutes.** Stir with whisk. **MW on High 3 minutes,** or until mixture starts to boil. Stir and let cool.

Slice bananas and stir carefully into cooled pudding. Put several vanilla wafers into each of four dessert bowls. Divide pudding mixture into bowls. Chill in refrigerator. Serve with whipped topping. Makes 4 servings.

CALCUTTA RICE PUDDING

In India, the true test for a bride is that her "Kheer" (Rice Pudding) meets the approval of her new mother-in-law!

2 eggs, well beaten
1⅓ cups lowfat milk
1 (5⅓-ounce) can evaporated milk (⅔ cup)
¼ cup sugar
¼ teaspoon nutmeg
½ cup quick-cooking (instant) rice
⅓ cup raisins
¼ cup toasted sliced almonds
1 (11-ounce) can Mandarin orange sections

Combine eggs, lowfat milk, evaporated milk, sugar and nutmeg in a 1½-quart round casserole. **MW on High 3 minutes,** stirring once. Blend in rice and raisins. **MW on High 3 minutes,** stirring once. Blend in almonds. Garnish top with orange sections. Serves 5 to 6.

NOTE: *May be served hot or cold.*

ICE CREAM

Begin making homemade ice cream in your microwave oven. The cooked base is then processed in your ice cream freezer.

Ice cream dates back to the time of Marco Polo, who told of the ices that the oriental people made. In the 1600's, a superb frozen "cream ice" was served in the royal courts. This delicacy instantly became a success and its popularity spread across the ocean to America in the 1700's. Our early presidents served ice cream at many state dinners.

The St. Louis Fair in 1904 is credited with the first ice cream cone. An ice cream vendor ran out of little serving dishes and a near-by waffle vendor came to the rescue by rolling his paper-thin waffles into cones to hold the ice cream.

Vanilla ice cream is the all-time favorite with many other flavors available as well. Commercial ice cream is homogenized and has stabilizers in it for uniformity and keeping quality. Homemade ice cream can be made by either a still-frozen or stirred method. In the still-frozen method, the base mix is poured into a flat tray to speed freezing. The tray is placed in the freezer and periodically the mixture is stirred to redistribute the ice crystals and to keep them small. This type of ice cream generally is not as smooth as that made in a crank-type freezer.

An ice cream freezer, either hand-crank or electric, is used to make stirred ice cream. Rock salt and ice are used to surround the container holding the base mix. The salt lowers the temperature of the ice and water bath for better freezing. The ice cream is constantly stirred until the dasher can no longer be turned.

To ensure yummy results:

• Use a whisk every two minutes in microwaving the base mix. Since we use High power to cook the ice cream mixture, a whisk will keep the ingredients well blended for uniform heating.

• Transfer the cooked base mixture to the metal ice cream container to chill before churning. It won't take as long to freeze the chilled base mix.

• Use rock salt rather than table salt which melts quickly. To harden, or ripen, ice cream, remove dasher and pack ice cream down into container. Cover ice cream with waxed paper or aluminum foil. Replace cover and ripen one to three hours before eating. Or, ice cream can be packed into plastic containers and placed in freezer to harden.

FRENCH VANILLA ICE CREAM

In honor of our ST. LOUIS POST DISPATCH

- 6 egg yolks
- 1½ quarts half and half, divided
- 1 cup sugar
- ½ pint heavy cream
- 1 tablespoon vanilla

Combine egg yolks and 1 quart of half and half in a 2-quart batter bowl. Add sugar. **MW on High 8 to 9 minutes,** whisking every 2 minutes, or until mixture coats a metal spoon. Chill in metal container. When chilled, add remaining 2 cups half and half, heavy cream and vanilla. Churn. Ripen. Yields 3 quarts.

JUDY'S CHOCOLATE ICE CREAM

It's the real thing!

- 1 (¼-ounce) envelope unflavored gelatin
- ¼ cup water
- 6 (1-ounce) squares unsweetened chocolate
- 1 quart half and half, divided
- 2 cups sugar
- 2 eggs, beaten
- 1 tablespoon vanilla

Soften gelatin in water. Place chocolate in a 2-quart batter bowl. **MW on 50% (Medium) 3 to 3½ minutes,** or until chocolate is melted. Blend 2 cups half and half and softened gelatin into chocolate. Stir in sugar. **MW on High 6 to 8 minutes,** whisking every 2 minutes, until mixture almost coats a metal spoon. Add a small amount of chocolate mixture to eggs; whisk egg mixture into chocolate mixture. **MW on High 2 to 3 minutes,** until mixture thickens slightly. Add remaining half and half and vanilla. Pour into metal ice cream freezer and chill. Churn. Ripen. Yields 3 quarts.

Variations: JUDY'S BROWNIE SPECIAL—Add small broken pieces of day-old brownies to chilled base mix before churning.

ROCKY ROAD—Add 2 cups miniature marshmallows and 1½ cups coarsely chopped nuts to chilled base mix before churning.

Ice Cream

FRESH FRUIT SWIRL

1½ quarts milk, divided
1 (13-ounce) can evaporated milk
2 cups sugar, divided
1½ tablespoons flour
3 eggs
1 teaspoon vanilla
3 cups crushed fresh fruit (strawberries, peaches, etc.)

Combine 1 quart milk and evaporated milk in a 2-quart batter bowl. **MW on High 7 to 9 minutes,** whisking every 2 minutes. Blend 1½ cups sugar and flour together; blend in eggs. Add a small amount of hot milk to thin egg mixture. Return sugar and egg mixture to hot milk. **MW on High 4 to 5 minutes,** or until mixture thickens slightly, whisking every 2 minutes. Add remaining 2 cups milk and vanilla; pour into metal container and chill.

Combine ½ cup sugar with crushed fruit; refrigerate.

Churn ice cream. When done, swirl fruit mixture throughout. Ripen. Yields 3 quarts.

CHEESECAKE ICE CREAM

3 egg yolks
1½ cups sugar, divided
1 pint half and half
3 (8-ounce) packages cream cheese
2 tablespoons lemon juice
2 teaspoons vanilla
1 pint plain yogurt
Favorite topping

Combine egg yolks and ½ cup sugar in a 4-cup glass measure. Whisk in half and half. **MW on High 5 to 5½ minutes,** whisking midway through cooking. Refrigerate.

Place cream cheese in mixer bowl and **MW on 50% (Medium) 3 to 3½ minutes.** Beat in remaining 1 cup sugar, lemon juice and vanilla. Add yogurt and chilled egg yolk mixture. Beat until smooth. Churn. Ripen. Yields 2½ quarts.

NOTE: *To serve ice cream, top with favorite topping. (We suggest strawberry or blueberry.) Or, pack churned ice cream into prepared graham cracker crust; freeze. When solid, top with favorite topping. To serve,* **MW on 30% (Medium-low) 1½ to 2 minutes** *for easier slicing.*

COOKIES, CAKES, PIES, CANDY, etc.

Quick Breads

EGG CARTON GEMS
Miniature Boston brown bread

- 2 foam egg cartons from large or extra-large eggs
- 1 cup buttermilk
- ¼ cup oil
- ⅓ cup dark molasses
- ½ cup flour
- ½ cup whole wheat flour
- ½ cup cornmeal
- 1 teaspoon soda
- ½ teaspoon salt
- ½ cup raisins
- ¼ to ½ cup chopped nuts

Spray each compartment of egg cartons with Pam, or line with miniature paper nut cups. Do not cut off lid, because it will be used as a cover.

Measure buttermilk into a 2-quart glass batter bowl. Add oil and molasses. Add flours, cornmeal, soda and salt. Stir until well combined. Fold in raisins and nuts.

Spoon batter into one egg carton, filling each compartment ⅔ full. Close lid. Place carton on a meat or bacon rack in oven and **MW on 50% (Medium) 3 minutes.** DO NOT PEEK! Leave lid closed and set carton flat on counter for 3 minutes. Meanwhile, fill second egg carton and microwave as before.

To remove "gems," turn carton upside down over a plate and loosen gently with the point of a knife. Spray empty cartons again with Pam, or line with nut cups. Repeat baking batter. Makes 36 to 48 gems.

NOTE: *We tried using one egg carton four times, but without the nut cups, the third and fourth batches stuck due to crumb residue.*

→ Quick Breads

CINNAMON BISCUIT RING

CiCi's son Richard likes to make this for the family on Sunday morning.

- 3 tablespoons margarine
- ⅓ cup packed light brown sugar
- ¼ cup raisins
- ¼ cup chopped nuts
- 1 tablespoon water
- ½ teaspoon cinnamon
- 1 (8-ounce) can refrigerated biscuits (10-count)

Place margarine, sugar, raisins, nuts, water and cinnamon in a 6-cup ring pan. **MW on High 1½ minutes,** or until mixture is bubbly. Cut each biscuit in half. Stir biscuits into sugar mixture, coating each piece. Arrange in dish. **MW on High 2½ to 3 minutes,** rotating dish midway through cooking. Invert ring on platter. Let stand, covered, 2 minutes before serving. Makes 4 to 5 servings.

MINCEMEAT COFFEE RING

Spread slices with orange-flavored cream cheese

- 1¾ cups flour
- 1½ teaspoons baking soda
- ½ teaspoon salt
- ¼ cup margarine
- ½ cup packed light brown sugar
- 2 eggs
- 1 cup prepared mincemeat

Grease a 6-cup ring mold with solid shortening. Dust with graham cracker crumbs.

Sift together flour, soda and salt. Set aside. Place margarine in a medium-size microwave-safe mixing bowl. **MW on High 45 seconds,** or until melted. Blend in sugar and eggs thoroughly. Add dry ingredients; blend. Stir in mincemeat and pour into prepared pan. **MW on 50% (Medium) 5 minutes,** rotating once or twice. **MW on High 2½ to 3½ minutes,** rotating once. Let stand 5 minutes. Invert on serving platter. Makes 8 servings.

Muffins

BASIC MUFFINS

1⅔ cups flour
½ cup sugar
2 teaspoons baking powder
½ teaspoon salt
⅛ teaspoon nutmeg or cinnamon
⅓ cup oil
¾ cup milk
1 egg

Optional toppings: chopped nuts, chopped granola, cinnamon-sugar mixture or toasted coconut

If using a microwave muffin pan with solid bottom, spray each compartment with Pam. For microwave muffin pans with steam vent holes in bottom, line with cupcake papers.

In a medium mixing bowl, combine flour, sugar, baking powder, salt and nutmeg. Blend together oil, milk and egg. Stir into dry ingredients just until moistened.

Fill compartments one-half full. Sprinkle optional topping on batter, if desired. **For 6 to 7 muffins, MW on High 2½ to 3 minutes,** rotating dish twice. Repeat procedure for remaining batter. Yields 14 muffins.

BACON-CHEESE MUFFINS

Microwave 7 slices bacon, crumble and set aside. Make Basic Muffins, reducing sugar and vegetable oil to ¼ cup each. Stir ½ cup shredded sharp cheddar cheese into batter. For topping, sprinkle crumbled bacon on top of each muffin. Proceed according to Basic Muffin directions. Yields 14 muffins.

Dear Microscope,
  How do you get the muffin pan into the microwave oven before the batter runs out?

Dear Microcook,
  For microwave muffin pans with steam vent holes in the bottom, line first with cupcake papers before filling with batter. If you don't have any cupcake papers, you'd better run fast after pouring the batter!

> **Muffins**

JAM-DANDY MUFFINS

Make Basic Muffins. **MW on High 2 to 2½ minutes.** Place 1 teaspoon jam and 2 teaspoons chopped nuts on top of each muffin, pressing into batter slightly. **MW on High 30 seconds.** Yields 14 muffins.

OATMEAL-RAISIN MUFFINS

Make Basic Muffins using 1 cup flour; ⅔ cup quick oats, uncooked; and ¼ cup brown sugar. Stir ½ cup raisins into batter. Select optional topping and proceed according to Basic Muffin directions. Yields 14 muffins.

STRAWBERRY MUFFINS

1¾ cups flour
⅓ cup sugar
2 teaspoons baking powder
½ teaspoon salt
⅓ cup oil
1 egg
1 (10-ounce) package frozen strawberries, thawed and divided

If using a microwave muffin pan with solid bottom, spray each compartment with Pam. For microwave muffin pans with steam vent holes in bottom, line with cupcake papers.

In a medium mixing bowl, combine flour, sugar, baking powder and salt. Blend together oil, egg and 1 cup strawberries and juice. Stir into dry ingredients just until moistened. Reserve remaining strawberries and juice to make Strawberry Butter, recipe below.

Fill compartments one-half full. For 6 to 7 muffins, **MW on High 2½ to 3 minutes,** rotating dish twice. Repeat procedure for remaining batter. Yields 14 muffins.

STRAWBERRY BUTTER

Combine ½ cup softened margarine and remaining strawberries and juice from Strawberry Muffins (approximately ¼ cup) in blender or food processor. Process until well blended. Refrigerate. Yields ¾ cup. *Use as spread for Strawberry Muffins.*

QUICK PUMPKIN GINGERBREAD

1 (14.5-ounce) package gingerbread mix
½ cup canned pumpkin
2 eggs
½ cup lukewarm water

Using solid shortening, grease a 6-cup microwave ring pan. Set aside. Combine gingerbread mix, pumpkin, eggs and water in a mixing bowl. Blend with a wooden spoon until well combined. Pour batter into prepared pan. Cover with wax paper. Rotating pan every 2 minutes, **MW on 70% (Medium-high) 7½ to 8½ minutes,** or until gingerbread tests done. Let stand flat on counter 5 minutes before turning out onto plate. Makes 12 servings.

"BASKET CASES"

Prepare batter for Quick Pumpkin Gingerbread above. Line four plastic pint-size baskets (such as those used for strawberries and cherry tomatoes) with wax paper. Fill baskets with batter and place on a microwave-safe tray. Rotating each basket one-half turn midway through cooking, **MW on 70% (Medium-high) 7½ to 8½ minutes,** or until gingerbread tests done. Let stand flat on counter 5 minutes before removing from wax paper.

To give as gifts, wrap individual loaves in plastic wrap, and decorate baskets as desired.

PUMPKIN BREAD

1 cup packed dark brown sugar
1 cup pumpkin
⅓ cup oil
2 eggs
1½ cups flour
½ teaspoon EACH cinnamon, nutmeg, baking soda & salt
¼ teaspoon baking powder
½ cup chopped nuts

Combine sugar and pumpkin in a medium mixing bowl. Add oil and eggs; blend thoroughly. Sift flour, cinnamon, nutmeg, baking soda, salt and baking powder together. Add to pumpkin mixture; mix well. Stir in chopped nuts.

Pour into a greased 9x5x3-inch glass loaf pan. Rotating several times as needed, **MW on High 9 to 10 minutes,** or until done. *(Bread will pull away from sides of utensil and look dry on bottom.)* Let stand 5 minutes. Turn out onto cooling rack. Yields 1 loaf.

CRANBERRY TORTE

1 cup cranberries
1 tablespoon lemon juice
1 (14-ounce) can sweetened condensed milk
⅓ cup margarine
2 cups graham cracker crumbs
¼ cup packed brown sugar
1 (8-ounce) container non-dairy whipped topping, thawed

Chop cranberries in food processor. Add lemon juice and condensed milk. Set aside.

Place margarine in an 8x8x2-inch glass dish, and **MW on High 1 minute.** Add graham cracker crumbs and sugar; blend well. Remove ¼ cup of crumb mixture and reserve for topping. Press remaining crumbs firmly in dish. **MW on High 2 minutes,** rotating dish once.

Pour cranberry mixture over crumb mixture. **MW on 70% (Medium-high) 5 to 6 minutes,** rotating dish once. When cooled to room temperature, swirl topping over cranberry layer. Sprinkle reserved crumb mixture on top. Refrigerate. Cut into squares. Makes 9 servings.

PRUNE DANISH SNACK CAKE

½ cup dried pitted prunes
⅓ cup oil
2 eggs
¾ cup sugar
⅓ cup buttermilk
1 teaspoon vanilla
1¼ cups Bisquick
½ teaspoon cinnamon
⅛ teaspoon allspice
½ cup chopped nuts

Insert steel blade in food processor; add prunes and oil. Pulse until prunes are chopped uniformly.

Transfer to mixing bowl and beat in eggs; then sugar. Add buttermilk and vanilla; blend. Stir in Bisquick, cinnamon and allspice. Blend in chopped nuts.

Prepare an 8x8x2-inch dish by lightly greasing it with solid shortening; sprinkle with sugar. Pour batter into prepared pan. **MW on 50% (Medium) 5 minutes, then on High 3 to 4 minutes,** rotating as needed. Cool. *Dust with powdered sugar if desired.* Makes 9 servings.

Bar Cookies

CHOCOLATE CHIP COOKIE BARS

- ¼ cup oil
- 1 cup packed dark brown sugar
- 2 eggs
- 1 teaspoon vanilla
- 1½ cups buttermilk baking mix (such as Bisquick)
- ½ cup chocolate chips
- ¼ cup chopped nuts

Measure oil into a 2-quart batter bowl. Stir in sugar, eggs and vanilla. Blend until all lumps of sugar dissolve. Stir in baking mix until batter is uniformly blended. Fold in chocolate chips and nuts.

Spray an 8x8x2-inch glass dish with Pam. Pour batter into dish. Rotating dish ¼ turn clockwise every 2 minutes, **MW on 70% (Medium-high) 9 to 10½ minutes.** Do not overbake; bars firm considerably after cooling. Cut into bars. Makes 16 bars.

6-MINUTE GRANOLA BARS

- ½ cup margarine
- 1 cup packed dark brown sugar
- 1 egg
- 1 teaspoon vanilla
- 1 cup flour
- 1 teaspoon cinnamon
- ½ teaspoon baking powder
- 2 cups raisin bran cereal
- 1½ cups quick oats, uncooked
- ½ cup wheat germ
- ½ cup chopped nuts

Shield corners of a 3-quart rectangular glass dish (see page 183). Spray with Pam; set aside.

Place margarine in a 2-quart glass bowl and **MW on High 1 minute,** or until melted. Blend in sugar, then egg and vanilla. Stir in flour, cinnamon and baking powder. Add cereal, oats, wheat germ and nuts; stir until well combined.

Using the back of a wooden spoon, press mixture evenly in prepared dish. Pack down as firmly as possible. Rotating dish every two minutes, **MW on High 6 to 7 minutes.** Do not overbake: granola bars firm considerably as they cool. Cut into bars and store in an airtight container. Makes 25 bars.

➡ *Bar Cookies*

GIANT COOKIE PIZZA

½ cup margarine (1 stick)
½ cup smooth peanut butter
½ cup sugar
½ cup packed dark brown sugar
1 egg
⅔ cup flour
¼ teaspoon soda
1 cup quick oats, uncooked
½ cup plain chocolate M&M's

Place margarine in a 2-quart glass batter bowl. **MW on 30% (Medium-low) 1 minute.** Blend in peanut butter and sugars until creamy. Stir in egg. Add flour, soda and oats. Fold in half of the M&M's and save the rest for the top of the "pizza."

Spray a 12-inch round glass pizza tray with Pam. Using fingers, pat the dough evenly out to edge of tray. **MW on 70% (Medium-high) 10 to 11 minutes,** turning tray ¼ turn every 2 minutes. One minute before "pizza" is finished cooking, sprinkle surface with rest of M&M's. Let cool on counter before serving. Use a pizza cutter and pie server to take up pieces. Makes 16 wedges.

COCONUT BAR COOKIES

6 tablespoons margarine
1½ cups graham cracker crumbs
1 (14-ounce) can sweetened condensed milk
1 (6-ounce) package chocolate chips
1 (3½-ounce) can flaked coconut (1⅓ cups)
¾ cup chopped nuts

Place margarine in a 2-quart rectangular glass dish. **MW on High 45 seconds,** or until melted. Stir in graham cracker crumbs and pat mixture evenly over bottom of dish. **MW on High 3 minutes,** rotating midway through cooking.

Pour milk over crumb bottom. Top evenly with chips, coconut and nuts, pressing down gently to level. **MW on 70% (Medium-high) 10 to 11 minutes,** rotating every 3 minutes. Cool before cutting. Yields 36 cookies.

NOTE: *If your oven has only high power, last step of recipe may be microwaved on High 7 minutes, rotating more often.*

CHOCOLATE MINT BROWNIES

  1 (15½-ounce) package brownie mix
  2 tablespoons water
  2 eggs
  ½ cup chopped nuts
 12 chocolate-covered Thin Mints (half of a 6-ounce box)

Using solid shortening, lightly grease bottom and sides of an (8x8x2-inch) glass dish. Dust with granulated sugar. Shield corners (see page 183). Set aside.

In mixing bowl, combine brownie mix, water and eggs; blend. Stir in nuts and pour mixture into prepared dish. **MW on 70% (Medium-high) 7 to 8 minutes,** rotating every 3 minutes. Place mints over top of brownies. **MW on 70% 2 to 3 minutes.** Swirl melted mints over top to make frosting. Cool before cutting. Yields 16 squares.

CARROT BROWNIES

  ½ cup margarine (1 stick)
  1 cup packed dark brown sugar
  2 eggs
  1 cup flour
  1 teaspoon baking powder
  1 teaspoon cinnamon
  ¼ teaspoon salt
  3 medium carrots, pared and grated finely (1½ cups)
  ½ cup chopped walnuts

Place margarine in a glass mixing bowl and **MW on High 1 minute,** or until melted. Blend in sugar; then eggs. Stir in flour, baking powder, cinnamon and salt. Fold in carrots and walnuts.

Spray an 8-inch square glass dish with Pam. Shield corners (see page 183). Pour batter in dish and level. **Rotating every 3 minutes, MW on 70% (Medium-high) 10 to 11 minutes,** or until brownies test done. Frost with Creamy Cheese Frosting, below. Yields 16 squares.

CREAMY CHEESE FROSTING

  1 (3-ounce) package cream cheese
  ¼ cup margarine
  1 tablespoon milk
  1 teaspoon vanilla
  2 cups powdered sugar, sifted

Place cream cheese and margarine in a glass mixing bowl. **MW on 50% (Medium) 1 to 1½ minutes,** or until softened. Blend in milk, vanilla and sugar. Beat until smooth. Frosts a one-layer cake.

➤ *Brownies*

FUDGE CAKE SQUARES WITH CHOCOLATE BAR FROSTING
A rich brownie from Switzerland

- ½ cup butter or margarine (1 stick)
- 3 tablespoons cocoa
- 1 cup sugar
- 2 large eggs
- 1 teaspoon vanilla
- ¾ cup flour
- 1 teaspoon baking powder
- 1 cup chopped nuts
- 1 (100-gram) Swiss or milk-chocolate bar without nuts (about 3½ ounces)

Using solid shortening, grease an 8-inch square glass baking dish. Do not flour. Shield corners (see below). Set aside.

Place butter in a glass mixing bowl and **MW on High 1 minute,** or until melted. Using a wooden spoon, stir in cocoa, sugar, eggs and vanilla. Add flour and baking powder and beat well. Blend in nuts. Pour batter into prepared pan and level.

Rotating every 3 minutes, **MW on 70% (Medium-high) 8 to 9 minutes,** or until cake tests done. Break chocolate bar into pieces and place evenly over surface of cake. Cover pan with plastic wrap (do not let it touch chocolate). Let stand 5 to 10 minutes. Remove plastic wrap and smooth melted chocolate over top of cake. Let cool before cutting. Yields 16 squares.

SHIELDING WITH FOIL

If the cooking pattern of your microwave oven is such that the corners of food overcook, shield baking dish. Cut four (2x1-inch) pieces of aluminum foil and tape around the outside corners of the dish using transparent or masking tape.

Thin areas or corners of foods cook more rapidly than thick areas or centers. Since metal reflects microwaves, we use small pieces of aluminum foil to "shield" areas of food which need less cooking. The general rules are:

- Use no more than one-fourth foil to three-fourths uncovered food area, so that microwaves can be absorbed.
- Use new pieces of foil and place the shiny side toward the food.
- Crimp the foil tightly to the surface, or use wooden toothpicks to secure it. Do not leave a corner or edge sticking out. This acts as an antenna, and could cause sparks.
- Do not let foil come in contact with the metal walls of your microwave oven—this can cause sparks. Shielding is also useful in defrosting thick foods in which the outer areas defrost and begin to cook before the center is defrosted. Use strips of foil to shield areas of food which begin to feel warm.

MEXICAN CHOCOLATE CAKE

Converted from our HOUSTON CHRONICLE Food Editor, Ann Criswell's book

- ½ cup margarine (1 stick)
- 2 (1-ounce) squares unsweetened baking chocolate
- ½ cup oil
- ¾ cup water
- ½ cup milk
- 1½ teaspoons vinegar
- 2 cups flour
- 2 cups sugar
- 2 eggs, beaten
- 1 teaspoon baking soda
- 1 teaspoon cinnamon
- 1 teaspoon vanilla

Place margarine and chocolate in a glass mixing bowl. **MW on High 2 minutes,** or until melted. Add oil and water; blend. To sour the milk, combine with vinegar. Add sour milk, flour, sugar, eggs, baking soda, cinnamon and vanilla to chocolate mixture. Stir well with a wooden spoon.

Using solid shortening, grease two (2-quart) 8x12-inch rectangular glass pans. Shield corners (see page 183). Pour half of batter into each pan. **Cooking one pan at a time, MW on 70% (Medium-high) 7½ to 8½ minutes,** rotating several times during baking. Leave cakes in pans. Cool slightly. Frost cakes while still warm using Mexican Chocolate Frosting, below. Makes 2 cakes, 8 servings each.

MEXICAN CHOCOLATE FROSTING

Also great on any cake

- ½ cup margarine (1 stick)
- 2 (1-ounce) squares unsweetened baking chocolate
- 6 tablespoons milk
- 1 pound powdered sugar
- 1 teaspoon vanilla
- ½ cup chopped pecans

Place margarine and chocolate in a glass mixing bowl. **MW on High 2 minutes,** or until melted. Add milk, sugar and vanilla. Beat until smooth. Stir in pecans. Frosts two layers.

DAPPER RABBIT CAKE

For debonair occasions, not only at Easter!

- 1 **(2-layer size) package cake mix***
- ⅓ **cup oil**
- 3 **eggs**
- **Water (subtract ¼ of amount called for on package)**
- **White frosting for a 2-layer cake**
- **Red food coloring**
- **Coconut**
- **Jelly beans**
- **Shoelace licorice**

Cut two circles of wax paper to fit bottom of an 8-inch round glass cake dish. Place one in dish and save one to use for second layer.

Sift cake mix into bowl. Using a wooden spoon, combine cake mix, oil, eggs and water. Pour half of batter into prepared dish. **MW on 50% (Medium) 5 minutes; then on High 1½ to 2 minutes,** rotating as needed. Turn this first layer out onto large serving tray to make face for rabbit. Remove wax paper after 5 minutes. Place the other circle of wax paper into same dish, and add remaining batter. Microwave second layer as above. When cool, cut second layer to make rabbit ears and bow tie as in diagram below.

Remove one fourth of frosting and tint pink to frost bow tie and lining of ears. Frost remaining areas of cake with white frosting and sprinkle with coconut. Use jelly beans for rabbit's nose and eyes, and licorice for whiskers.

*NOTE: *Light color cake mixes are easier to frost than chocolate ones, because any loose chocolate crumbs discolor the white frosting.*

Cakes

PINEAPPLE DOWNSIDE-UP CAKE

A scratch cake in less than 10 minutes

- ¼ cup margarine
- ⅓ cup packed light brown sugar
- 1 (16-ounce) can pineapple slices, drain and reserve juice
- 4 maraschino cherries, cut in half
- 1¼ cups flour
- ¾ cup sugar
- 2 teaspoons baking powder
- ½ teaspoon salt
- 1 egg
- ⅓ cup oil
- ½ cup reserved pineapple juice
- ½ teaspoon each: vanilla and almond flavoring

Cut two circles of waxed paper to fit bottom of 8-inch round layer cake pan. Place paper in pan.

Place margarine on top of waxed paper liner and **MW on High 45 seconds,** or until melted. Sprinkle brown sugar over margarine. Arrange 7 pineapple slices on top of sugar mixture. Decorate centers of slices with cherries.

Place flour, sugar, baking powder and salt in a medium mixing bowl. Blend together egg, oil, pineapple juice and flavorings. Pour into dry ingredients and use a wooden spoon to blend. Pour batter over pineapple slices. **MW on 50% (Medium) 5 minutes,** rotating dish once. **MW on High 4 to 4½ minutes,** rotating dish twice. When done, toothpick will come out clean when stuck in cake. Let cake stand in pan 5 minutes. Invert on serving platter. Makes 7 servings.

Quick-Tip ➡ To thaw an 8-ounce container of non-dairy whipped topping, **MW on 30% (Medium-low) 1 minute.**

Cakes

COCONUT GRIDIRON CAKE

Moist and delicious–decorated or not

- 1 (2 layer-size) white cake mix
- 2 eggs
- ⅓ cup oil
- Water*
- 1 cup milk
- ½ cup sugar
- 1 cup shredded coconut, divided
- 1 (8-ounce) carton frozen whipped topping, thawed

(*To determine amount of water, use ¼ less than your cake mix package directs.)

Sift cake mix into a large bowl. Add eggs, oil and water. Blend well, using a wooden spoon. Shield corners of a 2-quart rectangular glass dish (see page 183). Pour batter into dish. Rotating every 2 minutes, **MW on 50% (Medium) 8 minutes; then on High 3 to 4 minutes,** or until cake tests done. Set flat on counter.

Combine milk, sugar and ¼ cup coconut in a 2-cup glass measure. **MW on High 3 minutes,** or until mixture begins to simmer. Stir. Using a two-pronged meat fork, poke holes at ½-inch intervals all over cake, making sure to pierce completely through to bottom of dish. Pour milk mixture evenly over cake and let cool.

Spread whipped topping over cake and sprinkle with remaining coconut. Refrigerate several hours or overnight. Makes 12 servings.

NOTE: *To decorate as football "gridiron," tint remaining coconut with green food coloring before sprinkling over cake. Decorate with purchased plastic football players and goalposts. Can also be decorated for other sports.*

Cakes

LEMONADE CAKE

½ cup graham cracker crumbs
1 (6-ounce) can frozen lemonade concentrate
1 (2-layer size) yellow cake mix (without pudding in mix), sifted
1 (4-serving size) package instant vanilla pudding
½ cup oil
4 eggs

Using solid shortening, grease a 12-cup microwave Bundt pan. Sprinkle graham crackers into pan and tilt to coat. Set aside.

Transfer lemonade concentrate to a 1-cup glass measure. **MW on High 1 minute.** Add water, if necessary, to measure ¾ cup. In a mixing bowl, combine cake mix, pudding mix, oil, eggs and lemonade concentrate; blend well using a wooden spoon. Pour into prepared pan.

Set pan on an inverted pie plate on floor of microwave oven. (This is not needed in ovens with turntables.) **MW on 50% (Medium) 8 minutes; then on High 3 to 4 minutes,** rotating pan every 3 minutes. When cake tests done, set flat on counter, cover and let stand 10 minutes. Invert and cool. If desired, drizzle with Simple Glaze for Bundt Cakes, page 189. Makes 16 servings.

LEMON POPPYSEED CAKE

Finely-crushed graham crackers
1 (19-ounce) lemon cake mix
2 eggs
1 cup water
⅓ cup oil
¼ cup poppy seeds
1 teaspoon almond extract
Sifted powdered sugar

Using solid shortening, grease a 12-cup microwave Bundt pan. Dust with graham crackers; set aside.

Sift cake mix into a large bowl. Add eggs, water, oil, poppy seeds and extract. Blend using a wooden spoon. Pour into prepared Bundt pan. Turn a glass pie plate upside down on floor of microwave oven and place Bundt pan on it. (Not needed in ovens with turntables).

Rotating cake every few minutes, **MW on 50% (Medium) 8 minutes; then on High 2 to 3 minutes,** or until cake tests done. Let stand flat on counter 5 minutes. Loosen cake and turn out onto plate. Let cool and dust with powdered sugar, or drizzle with Simple Glaze for Bundt Cakes, page 189. Makes 16 servings.

TOURTA ATHINEIKI

Orange-walnut poundcake, Athens style

- 2 tablespoons crushed graham cracker crumbs
- 1 cup margarine (2 sticks)
- 2 cups sugar
- 4 eggs
- ½ cup orange juice
- Grated rind of 1 orange
- 1 teaspoon orange extract
- 3 cups flour
- 1 teaspoon baking powder
- ¼ teaspoon salt
- 1 cup walnuts, finely chopped

Using solid shortening, grease a 12-cup microwave Bundt pan. Add crumbs and tilt pan to coat evenly.

Place margarine in a 2-quart glass batter bowl. **MW on High 1 to 1½ minutes,** or until melted. Stir in sugar, then eggs, orange juice, rind and orange extract. Add flour, baking powder and salt, and stir until well blended. Fold in walnuts and pour batter into prepared pan. Let stand 10 minutes so that baking powder can begin to work.

Place a glass pie plate or microwave-safe tray upside down on floor of microwave oven. (Not needed in ovens with turntables.) Set cake pan on top. Rotating every few minutes or as needed, **MW on 50% (Medium) 8 minutes; then on High 3 to 4 minutes,** or until cake tests done. Let stand, covered, 5 minutes before turning out onto plate. Dust with powdered sugar or drizzle with Simple Glaze for Bundt Cakes, below. Makes 16 servings.

NOTE: *Delicious toasted or served with an orange sauce.*

SIMPLE GLAZE FOR BUNDT CAKES

- 2 tablespoons milk
- 1¼ cups powdered sugar

Mix milk and sugar together and drizzle over cooled cake.

Cakes

BLACK RUSSIAN CAKE

Bill Steiner's favorite. Liqueurs make this cake extra moist.

- 1 (2-layer size) devils food cake mix (without pudding in mix)
- 1 (4-serving size) package instant chocolate pudding
- ½ cup oil
- 4 eggs
- ½ cup hottest tap water
- 2 teaspoons instant coffee granules
- ¼ cup Kahlua (coffee liqueur)
- ¼ cup vodka
- ¼ cup creme de cacao
- 1 cup powdered sugar
- 2 tablespoons EACH strong coffee, Kahlua and creme de cacao

Using solid shortening, grease a 12-cup microwave Bundt pan. Set aside. Sift cake mix into a large mixing bowl. Add pudding mix, oil and eggs. Dissolve instant coffee granules in hot water and add to cake mix, along with Kahlua, vodka and creme de cacao. Blend using a wooden spoon. Pour into prepared pan. Turn a glass pie plate upside down on floor of microwave oven and place Bundt pan on it. (Not needed in ovens with turntables.)

Rotating cake every few minutes, **MW on 50% (Medium) 8 minutes; then on High 2 to 3 minutes,** or until cake tests done. Using a skewer or cake tester, punch holes through cake to bottom of Bundt pan at 1-inch intervals. Combine sugar with coffee, Kahlua and creme de cacao. Pour over cake in Bundt pan and let absorb before turning cake out onto plate. Makes 16 servings.

STREUSEL BUNDT CAKE MIXES

- 1 Pillsbury Streusel Bundt Cake Mix
- ⅓ cup oil
- Water (subtract ¼ of amount called for on package)
- Eggs (follow package amount)

Using solid shortening, grease a 12-cup microwave Bundt cake pan. Dust with graham cracker crumbs or finely-chopped nuts.

Sift cake mix into mixing bowl. Add oil, water and eggs; blend using a wooden spoon. Pour into prepared Bundt pan, layered with streusel mix according to package directions. **MW on 50% (Medium) 8 minutes; then on High 3 to 4 minutes,** rotating every three minutes. Let stand flat on counter 5 minutes before turning out onto plate. When cool, glaze according to package directions. Makes 16 servings.

NUTTY CRANBERRY CAKE

½ cup pecan meal or finely chopped pecans
1 (2-layer size) pudding-in-the-mix white cake mix, sifted
7 ounces water
⅓ cup oil
3 eggs
1 cup cranberries
½ cup chopped pecans

Using solid shortening, grease a 12-cup microwave Bundt pan. Sprinkle pecan meal into pan and tilt to coat.

Combine cake mix, water, oil and eggs; blend thoroughly. Finely chop cranberries in food processor, blender or food grinder. Fold cranberries and nuts into batter. Pour into prepared pan.

MW on 50% (Medium) 8 minutes; then on High 3½ to 4½ minutes, rotating pan as needed. When cake tests done, cover and let stand for 5 minutes. Invert on cake platter. Makes 12 to 16 servings. *Drizzle with Simple Glaze for Bundt Cakes, if desired (see page 189).*

MASQUERADE CAKE

Finely-crushed graham crackers
4 eggs, beaten
2 cups packed light brown sugar
½ cup oil
½ cup orange juice
1 teaspoon vanilla
2½ cups buttermilk baking mix (such as Bisquick), sifted
1 cup chopped pecans
1 (3½-ounce) can flaked coconut (1⅓ cups)

Using solid shortening, grease a 12-cup microwave Bundt pan. Dust with graham crackers; set aside.

In a large bowl, combine eggs, brown sugar, oil, juice and vanilla. Blend in sifted baking mix using a wooden spoon. Stir in pecans and 1 cup of coconut. Reserve remaining coconut for garnish. Pour into prepared pan.

Turn a glass pie plate upside down on floor of microwave oven and place Bundt pan on it. (Not needed in ovens with turntables.) Rotating cake every few minute, **MW on 50% (Medium) 8 minutes; then on High 3 to 4 minutes,** or until cake tests done. Let stand on counter 5 minutes before turning out onto plate. Makes 16 servings.

YULE LOG

An easy cake for all seasons

- 1 (16-ounce) pound cake
- 1 (12-ounce) package semi-sweet chocolate pieces
- 1 (3-ounce) package cream cheese
- 1 cup non-dairy whipped topping
- 2 tablespoons milk
- 1 tablespoon coffee liqueur, optional
- 1 cup powdered sugar, sifted
- ¼ cup sliced almonds
- 2 teaspoons margarine

Use a thin-bladed sharp knife or dental floss to trim the upper edge of cake, giving a more rounded appearance to resemble a log. *(Hold ends of a 12-inch length of dental floss in hands and use a sawing motion to slice.)* Slice cake horizontally to form three layers.

Place chocolate pieces in a 2-quart batter bowl and **MW on 50% (Medium) 2 to 2½ minutes.** Place cream cheese on top of chocolate. **MW on 50% 1¼ to 1½ minutes,** or until cream cheese and chocolate are softened. Blend thoroughly. Stir in whipped topping, milk and coffee liqueur. Stir in sugar.

To assemble cake, place bottom cake layer on serving tray. Spread with chocolate frosting. Repeat with remaining two layers. Frost all sides of cake with remaining frosting. Use tip of knife to form ridges in length of log.

Place nuts and margarine in a 1-cup glass measure. Stirring twice, **MW on High 1½ to 2 minutes,** or until golden. Sprinkle on top of cake. Refrigerate. Makes 6 to 8 servings.

PIE SHELLS

- All pie shells need to be pre-baked in the microwave oven before adding filling. This applies to fruit or custard-type pies, or quiches.
- Spray pie plate with Pam before placing pastry into it. This will prevent pastry from sticking to pie plate after microwaving.
- Use a clear glass pie plate to tell from the bottom if the pie shell is done. Look for a dry, opaque and blistered crust in a finished pie shell.
- If using a frozen pastry shell, pop it out of the foil liner while still frozen. It takes just a few minutes at room temperature to thaw. The pastry can then be molded to fit your pie plate.
- Prick pastry shell on the bottom and sides with a fork before microwaving (the same procedure as for a conventionally baked pie shell).
- **MW on High 4 to 5 minutes,** rotating plate midway through cooking. If baking a filling in pie shell, it does not need to cool first.

APPLE CRISP PIE

Use fresh apples or canned pie filling

- 2 pounds Rome apples
- ⅔ cup sugar
- ¼ cup flour
- 1 teaspoon cinnamon
- 1 (9-inch) pie shell, baked
- 4 tablespoons margarine
- ⅓ cup packed dark brown sugar
- ⅓ cup flour
- ⅓ cup oats
- ½ teaspoon cinnamon

Quarter, core and slice apples. Toss with sugar, flour and 1 teaspoon cinnamon. Pile into baked pie shell. Using a food processor or pastry blender, cut margarine into brown sugar and flour. Add oats and cinnamon and sprinkle over top of apples. Rotating once midway through cooking. **MW on High 9 to 11 minutes.** Cool slightly before cutting. Serves 6 to 8.

NOTE: *May also substitute 1 (20-ounce) can apple pie filling for apples, sugar and flour. Complete as above and* **MW on High 6 minutes.**

"I have a friend who has your microwave cookbook and thinks it's the greatest—hopefully you still have one left!"
Mrs. Roy Carmichael, Plainville, Kansas

FRESH PEACH PIE

The only hard part is peeling the peaches!

- 3 tablespoons water
- ¼ teaspoon almond extract
- 2 teaspoons ascorbic acid crystals (such as Fruit Fresh)
- 2 pounds firm ripe peaches (about 6)
- 3 tablespoons minute tapioca
- ¼ cup packed brown sugar
- ½ cup sugar
- ¼ teaspoon cinnamon
- ¼ teaspoon ginger
- 1 (9-inch) pie shell, baked
- 6 pecan shortbread cookies, crushed

Measure water, extract and ascorbic acid crystals into a 2-quart batter bowl; stir to dissolve. Peel peaches and slice uniformly. Add to mixture and coat peaches to prevent turning brown. Add tapioca, sugar and spices. Toss gently and cover with plastic wrap. **MW on High 6 minutes,** stirring once. Let stand 10 minutes before pouring into pie shell. Top with crushed cookies. Let cool 1 hour before slicing. Makes 6 servings.

PEACHES 'N CREAM TORTE

Baby food makes glazing easy

- 1 (8-ounce) package cream cheese
- ⅓ cup sugar
- 1 egg
- ¼ teaspoon almond extract
- ¼ teaspoon lemon extract
- 1 (9-inch) graham cracker crust, baked (see page 201)
- 3 fresh peaches, (about 1 pound)
- 1 teaspoon ascorbic acid crystals (such as Fruit Fresh)
- 1 (4¾-ounce) jar strained baby food Peach Cobbler
- 2 tablespoons sugar

Place cream cheese in small glass mixer bowl. **MW on 50% (Medium) 1½ minutes.** Beat cheese with electric mixer until fluffy. Add sugar, egg and extracts and beat until combined. Pour into baked crust and **MW on 70% (Medium-high) 3 to 3½ minutes.** Let cool and refrigerate.

Peel peaches and slice uniformly. To prevent peaches from turning brown, toss with Fruit Fresh and 2 tablespoons of water. Arrange over cream cheese layer. Combine baby food and sugar and pour over peaches as a glaze. Chill. Makes 8 servings.

PEACHY MINCEMEAT PIE

Less expensive than using canned mincemeat

- 1 (16-ounce) can sliced peaches, including liquid
- Water
- 1 (9-ounce) package condensed mincemeat
- 1 tablespoon brandy
- 1 (9-inch) pie shell, baked

Drain liquid from peaches into a 4-cup glass measure. Reserve peaches and add water to peach liquid to yield 1½ cups. Break mincemeat into small pieces and add to liquid. Stirring twice, **MW on High 3 minutes,** or until thickened. Add brandy and pour into pie shell. Arrange reserved peach slices on top of mincemeat in a pinwheel fashion. **MW on High 5 to 5½ minutes,** rotating once. Cool before slicing. Makes 6 to 8 servings.

PUMPKIN CHEESE PIE

- 1 (8-ounce) package cream cheese
- ½ cup packed brown sugar
- 3 eggs
- 1 (5⅓-ounce) can evaporated milk
- ½ teaspoon cinnamon
- ¼ teaspoon ginger
- ¼ teaspoon nutmeg
- ⅛ teaspoon allspice
- 1 cup cooked pumpkin
- 1 (10-inch) pie shell, baked
- 1 tablespoon margarine
- ½ cup chopped pecans

In a 2-quart batter bowl, place unwrapped cream cheese. **MW on 50% (Medium) 1½ minutes.** Blend in sugar; then add eggs, beating well. Add milk and use a wire whisk to thoroughly blend mixture. Add spices and pumpkin; blend. Pour mixture into pie shell and **MW on 70% (Medium-high) 10 to 11 minutes,** rotating dish twice. Center should jiggle slightly.

Place margarine in a 1-cup glass measure and **MW on High 30 seconds.** Stir in nuts and **MW on High 2 to 2¼ minutes,** stirring twice. Use a spoon to distribute hot nuts on top of warm pie. Cool before slicing. Makes 8 servings. *Garnish with whipped topping.*

FUDGE PECAN PIE

A favorite of CiCi's Pop, Ralph Cheney

- ¼ cup margarine
- 2 tablespoons flour
- ¼ cup cocoa
- ¾ cup sugar
- ¾ cup dark corn syrup
- 3 eggs
- 1 teaspoon vanilla
- 1 cup broken pecans
- 1 (9-inch) pie shell, baked

Place margarine in a 4-cup glass measure. **MW on High 40 seconds,** or until melted. Using a wire whisk, blend in flour and cocoa; then sugar. Measure corn syrup into a 2-cup glass measure. Beat eggs and vanilla into syrup.

Pour syrup mixture into cocoa mixture; blend well with whisk. Add pecans and pour into baked pie shell. **MW on 50% (Medium) 10 to 12 minutes,** rotating once midway through cooking. Cool to lukewarm before cutting. Makes 6 to 8 servings.

SWEET POTATO PECAN PIE

From our deep South JACKSON DAILY NEWS

- 1 cup mashed cooked sweet potatoes
- ¼ cup margarine, melted
- 3 eggs
- ¼ cup packed brown sugar
- Pinch of salt
- ¾ cup dark corn syrup
- 1 teaspoon vanilla
- 1 cup pecan pieces
- 1 (9-inch) pie shell, baked

Using blender or food processor with steel blade, combine potatoes, margarine and eggs. Add brown sugar, salt, corn syrup and vanilla and process until smooth. Add ⅔ cup of pecans and pulse to mix. Pour mixture into baked pie shell and sprinkle remaining pecans over top. **MW on 50% (Medium) 12 to 13 minutes,** or until center is almost set. Let cool at least 30 minutes before cutting. *Top with whipped cream or ice cream.* Makes 8 servings.

CHOCOLATE LOVERS' PIE

The favorite dessert of Rob Williamson, CiCi's oldest son.

- 2 (1-ounce) squares unsweetened chocolate
- 2 tablespoons margarine
- ⅓ cup flour
- 1½ cups milk
- ¾ cup packed brown sugar
- 2 eggs, lightly beaten
- ½ teaspoon vanilla
- 1 (9-inch) Chocolate Crumb Crust (page 201) OR pie shell, baked
- 2 cups whipped topping
- 1 tablespoon white creme de cacao liqueur (optional)

Place chocolate and margarine in a 4-cup glass measure. **MW on High 2 minutes,** or until chocolate melts. Using a wire whisk, blend in flour. Whisk in milk gradually. Blend in sugar, making sure that all lumps are dissolved.

MW on High 2 minutes; whisk. **MW on High 2 minutes,** or until mixture begins to thicken; whisk. Blend in eggs. **MW on High 1 minute,** or until thick. Whisk well and blend in vanilla. Pour into prepared pie shell. Let cool at room temperature; refrigerate.

Blend whipped topping and liqueur; spread over top of chilled pie. *Garnish with chocolate shavings, if desired.* Makes 6 to 8 servings.

LEMON MERINGUE PIE

- 1¼ cups sugar
- 3 tablespoons cornstarch
- ¼ cup fresh lemon juice
- 1 tablespoon grated lemon rind
- 3 eggs, room temperature, separated
- 1½ cups hottest tap water
- 1 (9-inch) pie shell, baked
- 6 tablespoons sugar
- 1 (9-inch) Pretzel Crust, recipe below

In a 4-cup glass measure, combine sugar and cornstarch. Add lemon juice, lemon rind, and egg yolks; blend. Using a whisk, blend in water. **MW on High 5 to 6 minutes,** whisking every two minutes, until mixture thickens. Pour into baked pie shell.

Preheat conventional or convection oven to 425°F. In small bowl of electric mixer, beat egg whites at highest speed until stiff. Gradually add 6 tablespoons sugar, one at a time. Spread meringue lightly over pie, making sure to seal meringue at edge of crust. **Bake in preheated oven 3 to 4 minutes,** or until browned as desired. Cool on a wire rack. Makes 6 servings.

PEACHY PRETZEL PIE

- 8 ounces cream cheese
- ½ cup sugar
- 1 (21-ounce) can peach pie filling
- 1 (1¼-ounce) packet whipped topping mix
- ½ cup milk
- 1 teaspoon vanilla
- 1 (9-inch) Pretzel Crust, recipe below

Place unwrapped cream cheese in a glass mixing bowl. **MW on 50% (Medium) 2 minutes,** stirring midway through. Blend sugar into softened cheese. Add pie filling; combine.

Using small mixer bowl, combine topping mix, milk and vanilla. Beat until mixture holds peaks. Fold into cheese mixture. Pour into cooled Pretzel Crust, see recipe below. Refrigerate 6 hours or overnight. If desired, decorate with pretzels or crushed pretzel crumbs. Makes 6 servings.

PRETZEL CRUST

- ½ cup margarine (1 stick)
- 3 tablespoons sugar
- Pretzels crushed to yield 1 cup crumbs

Place margarine in a 9-inch pie plate. **MW on High 1 minute,** or until melted. Add sugar and pretzel crumbs. Mix well. Using a spoon, pat mixture onto bottom and sides of plate. **MW on High 4 minutes,** rotating once midway through cooking. Cool.

CHOCOLATE COCONUT CRUST

- 2 (1-ounce) squares unsweetened chocolate
- 2 tablespoons margarine
- 2 tablespoons milk
- ⅔ cup confectioners sugar
- 1½ cups flaked coconut

Spray a 9-inch pie plate with Pam. Set aside. Place chocolate and margarine in a 1-quart microwave-safe bowl. **MW on 70% (Medium-high) 2 minutes,** or until melted. Stir in milk and sugar; mix well. Add coconut and toss until evenly mixed. Using a spoon, pat mixture onto bottom and sides of prepared pie plate. Refrigerate until firm (about one hour). Fill as desired.

COCONUT CREAM PIE

Hawaiian John Kelly's favorite

- 3 tablespoons cornstarch
- ½ cup sugar
- ⅛ teaspoon salt
- 2¼ cups whole milk
- 1 egg
- 2 teaspoons vanilla
- ½ cup flaked coconut
- 1 chocolate coconut crust (page 198), OR (9-inch) pie shell, baked
- Whipped topping

In a small bowl, combine cornstarch, sugar and salt. Measure milk into a 1-quart glass batter bowl. Using a wire whisk, blend in dry ingredients and then egg. Stirring with whisk every 2 minutes, **MW on High 6 to 7 minutes,** or until mixture just begins to boil. Add vanilla and cool to lukewarm. Stir in coconut and pour into crust. Refrigerate 4 hours or longer. Cover with whipped topping. Garnish with shredded coconut. Makes 6 to 8 servings.

KAHLUA CHIFFON PIE

- ⅓ cup Kahlua or coffee liqueur
- 1 (¼-ounce) envelope unflavored gelatin
- ½ cup milk
- 4 eggs, separated
- 1 cup sugar, divided
- 1 tablespoon unsweetened cocoa
- 1 (10-inch) pie shell, baked
- Grated chocolate

Combine Kahlua and gelatin; set aside. In a 4-cup glass measure, combine milk and egg yolks. Blend ½ cup sugar and cocoa together and add to milk mixture. Stirring midway through cooking, **MW on 70% (Medium-high) 3½ to 4 minutes,** or until slightly thickened. Stir in Kahlua and gelatin. Cool to room temperature.

Beat egg whites until soft peaks form. Gradually add remaining ½ cup sugar; beat until sugar is dissolved. Blend in Kahlua mixture. Pour into pie shell. Sprinkle top with chocolate. Refrigerate until firm. Makes 8 servings. *Garnish with whipped topping.*

Variation: RUM CHIFFON PIE - substitute rum for Kahlua in recipe above. Pour into a 10-inch Ginger Snap Crust.

PUMPKIN CHIFFON DELIGHT
Ann's favorite finale at Thanksgiving

- ¾ cup packed dark brown sugar
- 1 (¼-ounce) envelope unflavored gelatin
- 1 teaspoon cinnamon
- ½ teaspoon nutmeg
- ¼ teaspoon ginger
- ⅛ teaspoon cloves
- ⅛ teaspoon allspice
- ⅔ cup milk
- 3 eggs, separated
- 1½ cups pumpkin
- ¼ cup sugar
- 1 (10-inch) pie shell, baked
- Whipped cream

Blend together sugar, gelatin, cinnamon, nutmeg, ginger, cloves and allspice in a 2-quart batter bowl. Stir in milk, egg yolks and pumpkin; mix well. **MW on High 5 to 6 minutes,** or until mixture is heated through and gelatin and sugar are dissolved, stirring twice. Refrigerate until mixture mounds slightly when dropped from spoon.

Beat egg whites until almost stiff. Gradually add sugar and beat until sugar is dissolved. Fold pumpkin mixture into egg whites; blend. Pour mixture into baked pie shell and refrigerate 3 to 4 hours, or until firm. Top with whipped cream. Makes 8 servings.

MOONSHINE PIE

A Tennessee treat in honor of our KNOXVILLE NEWS-SENTINEL

- 1 cup packed dark brown sugar
- 5 tablespoons flour
- 1¼ cups milk
- 2 tablespoons margarine
- 2 large eggs, separated
- 1 tablespoon bourbon whiskey
- 1 (¼-ounce) packet unflavored gelatin
- ¼ teaspoon cream of tartar
- 1 Ginger Snap Crust, recipe below
- 2 tablespoons chopped pecans

Combine brown sugar and flour in a 4-cup glass measure. Using a wire whisk, blend in milk. Add margarine. Whisking midway through cooking, **MW on High 4 to 5 minutes,** or until mixture is a thick custard. Add egg yolks; blend.

In a small cup, combine whiskey and gelatin. Add mixture to hot custard. Set custard in ice water to cool, stirring occasionally.

Beat egg whites with cream of tartar until stiff. Fold into cooled custard mixture. Pour into cooled Ginger Snap Crust. Sprinkle with chopped pecans. Chill 4 hours. Makes 6 to 8 servings.

CRUMB CRUST

- 4 tablespoons margarine
- 1¼ cups graham cracker crumbs
- 2 tablespoons sugar

Place margarine in a 4-cup glass measure. **MW on High 45 seconds,** or until melted. Add graham cracker crumbs and sugar; blend well. Press crumb mixture onto bottom and sides of a 9-inch pie plate. **MW on High 2 minutes,** rotating plate once. Cool before adding filling. Yields 1 (9-inch) crumb crust.

> **Variation: GINGER SNAP CRUST** - Substitute crushed ginger snap cookies for graham cracker crumbs. *This is a great flavor teamed with lemon or rum filling.*

> **Variation: CHOCOLATE CRUMB CRUST** - Substitute crushed chocolate cookies for graham cracker crumbs.

Frozen Pies

FROZEN PUMPKIN SQUARES

- 4 tablespoons margarine
- 1¼ cups ginger snap crumbs
- 1 cup pumpkin
- ¼ cup packed brown sugar
- 1 teaspoon cinnamon
- ½ teaspoon nutmeg
- ⅛ teaspoon cloves
- ⅛ teaspoon allspice
- ½ cup pecan pieces
- ½ gallon vanilla ice cream

Place margarine in a 2-quart rectangular dish and **MW on High 45 seconds,** or until melted. Add ginger snap crumbs and mix thoroughly. Press crumb mixture in bottom only of dish. **MW on High 2 minutes,** rotating dish midway through cooking.

In a 2-quart batter bowl combine pumpkin, sugar, spices and nuts. Mix thoroughly. To soften ice cream, **MW on 30% (Medium-low) 3 minutes,** rotating carton twice. Blend ice cream into pumpkin mixture. Pour into ginger snap crust and freeze until solid. When ready to serve, let set at room temperature 5 minutes prior to cutting. *Garnish each square with a dollop of whipped cream and a pecan half.* Makes 12 servings.

PEANUT BUTTER PIE

Georgia peanuts in honor of our SAVANNAH NEWS-PRESS

- 1 (3-ounce) package cream cheese
- ½ cup crunchy peanut butter
- 1 cup sifted powdered sugar
- ½ cup milk
- 1 (8-ounce) carton frozen whipped topping
- 1 (9-inch) Peanut Pie Crust, page 203
- ¼ cup chopped roasted peanuts

Place unwrapped cream cheese in a 2-quart batter bowl. **MW on 50% (Medium) 1 minute.** Blend in peanut butter, then sugar. Add milk; blend until smooth. To thaw carton of whipped topping, **MW on 30% (Medium-low) 1 to 1½ minutes.** Fold topping into peanut butter mixture. Pile mixture into cooled Peanut Pie Crust. Garnish with peanuts. Freeze until firm. Remove from freezer 10 to 15 minutes before slicing and serving. Makes 8 servings.

> Frozen Pies

MUD PIE

For those who never grow up

- 1½ quarts coffee ice cream (coffee-almond-fudge, English toffee, or butter pecan are also good)
- 1 (9-inch) PECAN CRUST (recipe below)
- ½ cup chocolate fudge sauce
- 2 cups whipped topping or sweetened whipped cream
- 2 teaspoons instant coffee (powder, not freeze-dried)

To soften ice cream, **MW on 30% (Medium-low) 1 to 2 minutes.** Pack into prepared Pecan Crust. Drizzle with fudge sauce. Freeze until hard.

Blend topping with instant coffee. When ready to serve, spread topping over pie and garnish as desired. *Can also be frozen after topping.* Makes 6 servings.

PECAN CRUST

Great with other fillings, too.

- 6 tablespoons margarine
- ¼ cup packed brown sugar
- ¾ cup pecan meal or finely-chopped pecans
- ¾ cup flour

Spray a 9-inch microwave-safe pie plate with Pam. Set aside. Place margarine in a mixing bowl. **MW on High 1 minute,** or until melted. Stir in sugar, then pecans and flour. Blend well. Using a spoon, pat mixture onto bottom and sides of prepared pie plate. **MW on High 4 minutes,** rotating once after 2 minutes. Let cool before filling.

PEANUT PIE CRUST

- ½ cup margarine (1 stick)
- 1 cup flour
- ½ cup finely-chopped roasted peanuts

In mixing bowl, or using food processor, cut margarine into flour to a fine crumbly mixture. Stir in peanuts. Spray a 9-inch glass pie plate with Pam and press mixture evenly into it. **MW on High 5½ to 6½ minutes,** rotating plate twice. Cool.

Candy

Dear MicroScope,

I can't seem to melt chocolate in the microwave without scorching it. What am I doing wrong?

Dear MicroCook,

Chocolate must be melted on 50% power (Medium). Remember that chocolate continues to hold its shape in the microwave even though it is melted. You may look through the window of the oven, and the chocolate may still look solid. But, open the door and stir it. It's probably softer than you think it is!

Even though you are using the correct power level, it is still possible to scorch chocolate if you microwave it too long. Use the following times for melting chocolate, stirring midway through.

CHOCOLATE MELTING CHART

AMOUNT	TIME ON 50% (MEDIUM)
1 cup chocolate chips (6-ounce bag)	2½ to 3 minutes
2 cups chocolate chips (12-ounce bag)	5 to 6 minutes
1 square baking chocolate (1 ounce)	1½ to 2 minutes

BUCKEYE CANDY
In honor of our HAMILTON (Ohio) JOURNAL NEWS and TOLEDO BLADE

- 1 cup margarine
- 1½ cups graham cracker crumbs
- 1 (1-pound) box powdered sugar
- 1½ cups chunky peanut butter
- 18 ounces semi-sweet chocolate chips
- 2 tablespoons paraffin shavings

Place margarine in a 2-cup glass measure and **MW on High 2 minutes,** or until melted. Combine graham cracker crumbs and powdered sugar in food processor; blend. Add peanut butter and pulse to blend with dry mixture. Pour in melted margarine and continue to pulse until mixture is well blended. Shape balls approximately 1-inch in diameter.

Combine chocolate chips and paraffin in a 4-cup glass measure and **MW on 50% (Medium) 7½ to 8½ minutes,** or until chocolate is melted and very warm. Use a bamboo skewer or a round wooden toothpick to pierce and pick-up each "buckeye" to dip in melted chocolate. Do not completely dunk but leave an area approximately ½-inch in diameter of the peanut butter ball exposed to resemble a buckeye (a nut from Buckeye trees). Place on wax paper-lined cookie sheets. Refrigerate until set. Yields approximately 85 pieces.

Candy

NEVER-FAIL FUDGE

Our all-time favorite candy from CiCi's mom, Carolyn Cheney; mails well.

- 2¼ cups sugar
- ¼ cup margarine
- ⅔ cup evaporated milk
- 1 (7-ounce) jar marshmallow creme
- 1 (6-ounce) package semisweet chocolate chips
- 1 square baking chocolate (1-ounce)
- ½ cup chopped pecans or walnuts

Combine sugar, margarine, milk and marshmallow creme in a 2-quart glass batter bowl. **MW on High 3 minutes;** stir well. Continue to **MW on High 2 to 3 minutes,** or until mixture boils. **Then MW on 50% (Medium) and boil 5 minutes.** Reduce power if candy starts to boil over. Mixture should be a light almond color. Add chocolate chips and chocolate; stir until melted. Fold in nuts and turn mixture into a buttered 8-inch square pan. Cool before cutting into squares. Yields 2 pounds.

EASY ROCKY ROAD FUDGE

- 1 (6-ounce) package semi-sweet chocolate chips
- 1 (6-ounce) package butterscotch flavored chips
- 1 (14-ounce) can sweetened condensed milk
- 1 cup walnuts, coarsely chopped
- 2 cups miniature marshmallows

Pour packages of chips into a 2-quart batter bowl. **MW on 50% (Medium) 5 minutes,** or until chips are melted. Stir in milk. Add nuts and marshmallows. Turn mixture onto a foil-lined tray and shape into a heart for Valentine's Day, a wreath for Christmas, etc. Or, pour mixture into a foil-lined heart-shaped pan. Refrigerate 1 to 2 hours, or until firm, before removing and cutting. Yields approximately 2 pounds.

Quick-Tip ➡ Save money! Make your own Homemade Sweetened Condensed Milk. See recipe on page 182 in our *Microwave Know-How* book, *MicroScope Savoir Faire*. (See coupon in back for ordering information.)

Candy

PEANUT BUTTER CANDY

Imitation of Reese's Peanut Butter Cups

- 1 cup margarine (2 sticks)
- 1¼ cups crunchy peanut butter (one 12-ounce jar)
- 1 (1-pound) box powdered sugar
- 2½ cups graham cracker crumbs (almost half of a 16-ounce box of graham crackers, crushed)
- 1 (6-ounce) package semisweet chocolate chips

Spray a (9x13x2-inch) metal or glass sheet cake pan with Pam; set aside. Place margarine in a 2-quart casserole. **MW on High 1½ minutes,** or until melted. Stir in peanut butter until well blended. Stir in sugar and crumbs. Pat mixture evenly on bottom of prepared pan.

Place chips in a 4-cup glass measure. **MW on 50% (Medium) 2½ to 3 minutes,** or until melted. Spread over top of peanut butter mixture. Refrigerate until hard. Cut into squares and keep in an airtight container at room temperature. Yields 64 pieces.

MOCHA BRANDY CANDY

Melts in your mouth

- 1 (12-ounce) package semisweet chocolate chips
- 1 cup soft butter or margarine
- 2 eggs
- 4 tablespoons brandy
- 1 tablespoon instant coffee powder or crystals
- 2 teaspoons vanilla
- 2 cups powdered sugar
- 1½ cups finely-chopped pecans

Place chocolate chips into a glass mixing bowl and **MW on 50% (Medium) 5 minutes.** Stir and add margarine. **MW on 50% 1 to 2 minutes,** or until mixture can be blended. Beat in eggs.

Place brandy in a 1-cup glass measure. **MW on High 30 to 45 seconds.** Dissolve coffee powder in brandy. Add to chocolate mixture, along with vanilla. Blend in powdered sugar. Chill several hours.

When mixture is firm enough to handle, shape into small balls and roll in pecans. Chill well before serving. Yields 6 dozen.

WHITE CHOCOLATE PRETZELS

8 ounces white chocolate
1 tablespoon paraffin shavings
1 (10-ounce) box thin pretzels (not sticks)
Green or red sugar sprinkles, optional

NOTE: *White chocolate can be purchased in stores specializing in nuts and candies.*

Put chocolate into a 1-quart batter bowl. **MW on 50% (Medium) 2½ to 3 minutes,** stirring midway through melting. Add paraffin and stir until it melts. Dip one side of pretzels in melted chocolate and place on waxed paper with the uncoated side down. Decorate with sprinkles if desired. Let dry and store in airtight container. Yields 36.

CHOCOLATE-COVERED EASTER EGGS

½ cup butter or margarine
1 pound box confectioners' sugar
Flavoring (choose one of the following):
    For Vanilla Buttercream Filling, add 1 teaspoon vanilla and 1½ tablespoons cream
    For Orange Filling, add 1 teaspoon orange extract and 2 tablespoons orange juice
    For Cherry Filling, add 2 tablespoons maraschino cherry juice and 2 tablespoons chopped maraschino cherries.
1 (12-ounce) package real chocolate chips
¼ cup paraffin shavings

Place butter in a glass mixing bowl and **MW on 30% (Medium-low) 1 minute,** or until softened. Add sugar and your choice of flavoring. Mix with fork or use a food processor to blend well. Using the palms of your hands, shape mixture into 100 egg shapes, each about 1-inch long. Place on a wax-paper covered cookie sheet and freeze.

Place chocolate chips in a 4-cup glass mixing bowl. **MW on 50% (Medium) 5 minutes,** or until melted. To shave block of paraffin, use a vegetable peeler. Measure ¼ cup shavings into melted chocolate chips and stir until melted.

Remove about 25 egg shapes at a time from freezer. Dip each egg into melted chocolate using two forks. Turn egg over so that entire surface gets covered with chocolate. Remove with forks to a piece of waxed paper. If chocolate mixture cools and gets too thick to coat eggs, **MW on 50% until melted.** Repeat with remaining eggs. Let stand at room temperature until dry and firm. Store in an airtight container. Yields 100 chocolate covered Easter eggs.

Candy

PEANUT BRITTLE

- ½ cup light corn syrup
- 1 cup sugar
- 1 cup raw peanuts
- 1 tablespoon margarine
- 1 teaspoon vanilla
- 1 teaspoon soda

Cover a cookie sheet with aluminum foil; set aside. Measure corn syrup into a 2-quart glass batter bowl. Add sugar and peanuts. **MW on High 4 minutes;** stir. **MW on High 4 minutes;** add margarine and vanilla. **MW on High 1 minute.** Stir in soda and spread on prepared cookie sheet. Cool and break into pieces. Yields 1 pound.

Variation: PECAN BRITTLE. Substitute 1½ cups pecan pieces and ⅛ teaspoon salt for raw peanuts in recipe above.

RIM ROCK CRUNCH

Specialty of Methodist choir camp

- 1 cup light corn syrup
- 1 cup sugar
- 1 cup peanut butter
- 1 (20-ounce) box raisin bran

Combine corn syrup and sugar in a 4-cup glass measure. **MW on High 3 minutes,** or until mixture is dissolved, stirring once. Stir in peanut butter until completely blended.

Place raisin bran in a 6-quart container and pour sauce over top; blend thoroughly. Pat mixture onto a large buttered cookie sheet and refrigerate. When completely chilled, crunch can be broken into pieces or cut into squares and stored between sheets of waxed paper in an air-tight container. In warm climate, keep refrigerated. Yields 2 quarts.

"I've had my microwave for 5 years. But I've used it more since my neighbor has been sharing her cookbook with me. It is wonderful! I feel we can really trust the recipes to turn out delicious, because, so far, everything I've tried has been a success with my family."

Cynthia Gontarek, Groves, Texas

➡ *Preserves*

CRANBERRY JALAPEÑO JELLY

Red instead of the usual green

- 6 jalapeño peppers
- 2½ cups cranberry juice cocktail
- 7 cups granulated sugar
- 6 ounces liquid fruit pectin
- 1 cup vinegar
- Red food coloring, optional

Wearing lightweight rubber gloves, quarter and remove seeds from peppers. Place peppers and cranberry juice in blender and process until peppers are very finely chopped. Combine with sugar in a 4-quart simmer pot. Cover with plastic wrap (so that you can see mixture as it cooks). **MW on High 18 to 20 minutes,** or until mixture comes to a full rolling boil. Add pectin and **MW on High 2 to 3 minutes,** or until mixture returns to a boil. Add vinegar and food coloring, if desired. Pour mixture into sterilized jelly glasses and seal. Yields 8 (8-ounce) jars.

Serving Suggestion: *Spread ½ cup of jelly over an 8-ounce block of cream cheese. Serve with assorted crackers.*

STRAWBERRY PLATTER JAM

- 1 quart strawberries
- 3½ cups sugar
- 1 (1¾-ounce) package powdered pectin
- ½ cup water

Wash and hull strawberries; leave whole. Combine berries, sugar, pectin and water in a 4-quart simmer pot. Cover with plastic wrap. Stirring midway through cooking. **MW on High 7½ to 8 minutes,** or until sugar is completely dissolved. Stir, re-cover and continue to **MW on High until a full rolling boil is reached.** Reduce power level as necessary to prevent a boil-over and **MW 5 minutes.** Pour mixture into a 3-quart rectangular dish and let stand 1 hour or until beginning to set. Ladle into sterilized jars; cover. Let stand at room temperature until completely set. Store in refrigerator. Yields 4 to 5 (8-ounce) jars.

Preserves

SUPER-EASY STRAWBERRY PRESERVES

So fresh tasting and homemade

- 2 (10-ounce) packages frozen strawberries
- 3 tablespoons powdered fruit pectin
- 2 cups granulated sugar
- 1 tablespoon lemon juice

Place frozen strawberries in a 2-quart batter bowl or casserole. **MW on High 4 minutes,** breaking up frozen chunk after 2 minutes. Stir in pectin and **MW on High 2 minutes.** Stir in sugar and lemon juice. **MW on High 6 minutes,** stirring every 2 minutes. Pour into glass jars, cover and refrigerate when cooled to room temperature. *Can be thinned for use as an ice cream or cake topping.* Yields 4 cups.

TEXAS CAVIAR

A specialty from Neiman-Marcus

- ¼ cup red wine vinegar
- 1 clove garlic, minced
- ¼ cup thinly-sliced onion
- ½ teaspoon salt
- ¼ teaspoon pepper
- 1 cup oil
- 3 (16-ounce) cans blackeyed peas, drained

Measure vinegar into a 2-cup glass measure. Add garlic, onion, salt and pepper. Cover and **MW on High 2 minutes.** Measure oil into same cup and combine well. Pour mixture over blackeyed peas; stir. Store in a glass jar in refrigerator two weeks so that flavors can blend. Transfer to decorative jars to give as gifts. Yields 1 quart.

CRANBERRY ORANGE RELISH

Unbeatable flavor combination

- 2 oranges, quartered and seeded
- 1 pound fresh cranberries, washed
- 2 cups sugar

Chop oranges in food processor or blender. Combine with cranberries and sugar in a 3-quart casserole. Cover and **MW on High 15 minutes,** stirring every 5 minutes. Let stand, covered, 10 minutes. Cool and store in refrigerator. Yields 1 quart.

BABY FOOD

If there's a baby in your family, you've probably discovered how a microwave oven can be a lifesaver. Even if you're half asleep in the middle of the night, you can zap a bottle into the microwave oven and quickly quell the crying!

We want to encourage you to use microwaves for more than just reheating baby's food. Try making your own baby food! It is preferable to those commercially prepared, both economically and nutritionally.

According to *Consumer Reports*, "Commercial baby food isn't as nutritious as it could be. And the starches, sugar, and salt commonly added to baby food aren't necessarily good for infants."

Results of a long-term survey of infant nutrition were published in the March, 1982, *Journal of the American Dietetic Association.* The research concluded that "Mothers should not feed infants canned products that contain added salt, or add salt to homemade foods. Milk and milk products alone provide enough sodium for infants up to 18 months of age." Salt has been cited as a possible contributing factor to hypertension.

Added sugar serves no good nutritional purpose. It is empty calories which can lead to tooth decay, even in babies. Some nutritionalists now blame sugar in baby foods for stimulating a craving for sweets that may continue throughout life.

Other criticisms of commercially-prepared baby food includes:

1. Modified starches are added as thickening and to prevent ingredients from separating. These are empty calories.

2. Due to heat processing, commercial baby food is especially low in the vitamin thiamine.

3. Chemical preservatives and flavor enhancers are not needed by babies.

Consumer Reports concludes, "Homemade baby foods are more nutritious, ounce for ounce, primarily because the added and unnecessary ingredients in commercial foods take up a lot of room. Homemade foods contain a higher percentage of protein than similar commercial foods." They also cost only one-fourth to one-half the price of commercial baby foods, depending on the variety.

With a microwave oven, a blender or food processor, and a freezer, you can be in your own baby's food business! As your pediatrician recommends adding varieties of foods to your baby's diet, prepare them in your microwave oven as you would for the family. Do not add salt, sugar or flavorings. Add liquid for desired consistency, and puree in blender or food processor.

Baby Food

There are two methods of freezing baby's food. For "The Cube Method," pour pureed food into plastic ice cube trays. Freeze quickly and then "pop" out of trays into freezer-weight plastic bags for storage.

For "The Blob Method," drop blobs of pureed food onto a foil-lined cookie sheet or shallow pan. Freeze quickly and "pop" blobs off foil into freezer-weight plastic bags. Seal, label and date bags of "cubes" or "blobs." Food retains optimum quality for two months.

To feed a hungry baby fast, place frozen food in a microwave-safe dish and microwave on high power approximately one minute, or until warm. (You don't need to defrost frozen food on reduced power first.) For several different foods, place in a divided dish and microwave on high power until warm, stirring when possible to promote even heating.

Use your temperature probe to reheat bottles and refrigerated baby foods. Set for 110° on 70-percent power.

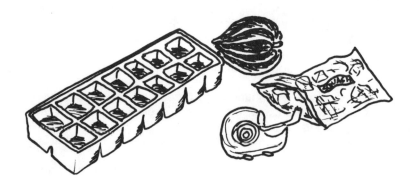

Dear MicroScope,

On the Playtex boxes of disposable baby nursers, it says, "Do not heat in microwave oven." Why not?

Dear MicroCook,

We called the Playtex Corporation in Dover, Delaware, to find out. Their spokeswoman told us that since microwaving time required to heat the "bottles" varies from oven to oven, the manufacturer declined to put a timing chart on the boxes. There is absolutely no danger to either the milk, the baby, or the plastic components. The disposable bags, plastic holder and nipple are all microwave safe. We recommend microwaving milk products on 70-percent power or lower to prevent curdling or boiling over.

DELICIOUS APPLESAUCE

3 Delicious apples, cored and quartered (about 1 pound)
3 tablespoons water

Place apples and water in a 1-quart casserole; cover. **MW on High 3 to 4 minutes,** or until fork tender.

Transfer mixture to food processor or blender. Puree until smooth. Yields 12 (1-ounce) cubes or blobs.

PRETTY PEAS

1 (10-ounce) box frozen peas
½ cup milk OR ⅓ cup water

Place box of frozen peas on a paper towel or plate. **MW on High 5½ to 6 minutes.**

Empty contents of box into food processor or blender. Add milk or water and puree until smooth. Yields 12 (1-ounce) cubes or blobs.

ACORN SQUASH

1 acorn squash (1 pound)
3 to 4 tablespoons natural apple juice

Pierce squash with an icepick so that steam can escape. Wrap in waxed paper to promote even-cooking. **MW on High 7 to 8 minutes.** Let cool slightly. Cut in half, remove seeds and scrape squash pulp into food processor or blender. Add apple juice and puree until smooth. Yields 14 (1-ounce) cubes or blobs.

BABY'S FIRST MEATS

½ pound ground lamb, beef or pork
½ cup water

Crumble meat into a 1-quart glass batter bowl. Add water. **MW on High 4 to 4½ minutes,** breaking up meat midway through cooking. Let cool until fat congeals on surface.

Skim fat and discard. Transfer cooked meat and liquid into food processor or blender. Puree until smooth. Yields 10 (1-ounce) cubes.

NOTE: *We found that by cooking the meat in liquid, it remained softer with few lumps; thus was most satisfactorily processed.*

Herb Drying ←▬▬▬▬▬▬▬▬▬▬▬▬▬▬▬▬▬

HERB DRYING

"Parsley, sage, rosemary and thyme." Yes, you can dry herbs in the microwave oven, with beautiful colors and quick results.

Growing fresh herbs at home is becoming increasingly popular. They are sometimes available in large supermarkets, and you can always find parsley. Drying herbs is the best way of preserving them for use after the growing season is past.

To seek advice from the experts, we visited Madalene Hill and her daughter Gwen Barkley of Hilltop Herb Farm in Cleveland, Texas. More than 2000 varieties are grown there and shipped to all states in the Union and to many foreign countries.

We were surprised to learn that "herb," like the man's name, Herb, with the hard "h," is the preferred pronunciation by "herbalists," also pronounced with a hard "h!" Madalene jokingly told us, "Herbalists don't trust people who say '-erb'!"

Madalene and Gwen shared some of their vast knowledge of herbs with us. During the growing season, cut herbs every three days. Don't be afraid, because it is good for them. When using fresh herbs, use two to three times the amount of dried herbs. Don't be afraid to experiment when seasoning foods with fresh herbs, because it's difficult to add too much.

All herbs are compatible with any foods, when used in the proper amounts.

"Herbs don't grow well on windowsills," Madalene told us. She graciously took us into the gardens and snipped samples for testing in our microwave oven. These are our findings.

MICROWAVE HERB DRYING

1. Cut sprigs of herbs early in the day when their aromatic oil content is highest.
2. Do not remove leaves from their stems. They dry more uniformly and are easier to handle this way.
3. Wash herbs and dry thoroughly. Let them air-dry in the kitchen several hours. If any moisture remains on the herbs, they will "cook" and not dehydrate.
4. Place three paper towels on a glass pie plate and arrange on the top layer five sprigs of herbs (or about ½ cup) in a wreath configuration. Cover with one paper towel.
5. Microwave on high power 3 to 3½ minutes, rotating plate midway through drying time.
6. Remove paper towel and let dried herbs cool. Store in an airtight jar in a cool place (refrigerator or freezer best).
7. Here is a suggestion from Sam Beneski, our microwaving friend in Austin, Texas: "If you have dried a large quantity of one kind of herb, keep the majority of it in a large jar, and transfer some to a smaller jar as needed. By not opening the large jar all the time, the flavor will be retained longer."

IMPORTANT NOTE: *Since herbs have a low moisture content, only microwave a dozen batches consecutively, and then let the oven rest. Alternate two pie plates, because the one in use will get hot as herbs dehydrate.*

➡ *Flower Drying*

FLOWER DRYING

Not only can you cook delicious foods in your microwave oven; you can also use it to dry flowers. Spring's abundance of beautiful flowers inspired us to experiment in flower drying.

Gather together the following equipment and you're ready to get started drying flowers the microwave way.
- Plastic gloves (thin disposable type)
- Drying medium: Silica Gel (available at craft/hobby shops) or cat litter (non-chlorophyll type)
- Containers (unwaxed paper cups, cardboard shoe box, glass custard cups or bowls)
- Scissors
- Fresh flowers
- Toothpicks
- Soft paint brush
- 20-gauge florist wire and green floral tape
- Clear craft spray

We suggest that you experiment with specimens that are not your 1st place blue ribbon winner or your daughter's bridal bouquet until you get a feel for drying flowers in the microwave.

Following are some of the tips we discovered as we microwaved our bouquets. See General Directions on page 216.
- Protect your hands by wearing thin disposable plastic gloves. Not only does the drying medium dry the flowers; it also dries your hands. The heavier lined rubber gloves are too cumbersome to work with the flowers easily.
- Cat litter is considerably less expensive (75¢ for a 5-pound bag) than Silica Gel ($6 for 1½ pounds). Thick flowers such as carnations, rose buds and chrysanthemums do best in the Silica Gel. Thin flowers such as wild flowers (from your own property–not those along the road!) and pansies did fine in the cat litter.
- Choose flowers or leaves that are fresh and in their prime. Those that are just beyond their best will continue to deteriorate after drying. The bright-colored flowers seem to do very well; however, color change takes place in some flowers. Don't try to dry flowers that still have morning dew on them.

Dear Microscope,
    Can you dehydrate fruits and vegetables in a microwave oven?

Dear Microcook
    NO. To dehydrate foods, hot air must be used to evaporate moisture without cooking the food. In a microwave oven, foods cook with little moisture loss since the oven contains no heat. Microwave/convection combination ovens can be used to dehydrate foods. However, the best appliance for quantity dehydrating is a dehydrator.

Flower Drying

GENERAL DIRECTIONS FOR DRYING FLOWERS

1) Leaving ½ to 1-inch attached to flower, clip off remainder of stem.
2) Partially fill container (paper cup, glass custard cup, etc.) with drying medium. Place flower, stem down, in medium. Use a toothpick to lift and separate petals to surround and cover the flower completely with drying medium. A shoebox serves as a nice tray for several containers.
3) Place a 1-cup glass measure of water to the rear of the microwave and place containers with flowers in front of water. This prevents the flowers from over-drying and falling apart.
4) Microwave on High. There are charts in several microwave cookbooks to serve as guidelines for particular flowers. We found that we had to get a "feel" for timing but generally thin flowers such as wildflowers take 90 seconds for 4 to 6 containers. Two medium-sized chrysanthemums (2-inch diameter) take approximately 2½ to 3 minutes.
5) Let flowers stand in drying medium 8 hours. Gently pour off drying agent and use a soft brush to remove particles that cling to petals. Save the drying agent for repeated use.
6) Insert florist wire through stem and secure with green floral tape. Spray finished flowers with clear craft spray to help hold their color and make them more lasting.

GENERAL DIRECTIONS FOR DRYING LEAVES

We tried drying leaves in each of the drying mediums as well as between paper toweling. The color was much better using paper toweling. Wash leaves (either individual or several on a stem) and shake off excess water. Sandwich leaves between 6 thicknesses of paper toweling (3 on top and 3 beneath). Place a 1-cup glass measure of water to the rear of the microwave and place leaves between paper toweling in front of water. Microwave on High approximately 1½ minutes; flip over top to bottom and microwave an additional 1½ minutes. Timing may need to be adjusted depending up amount of leaves being dried.

For charts to serve as a guideline to timing the drying of flowers and leaves in the microwave, you might look at:

Tout de Suite a la Microwave, Volume II, by Jean Durkee or

The New Magic of Microwave Cookbook, by Magic Chef.

INDEX

Almond Chicken Dip 23
American Mah-Jong 68
Ann Landers' Meat Loaf 84
Annapolis Stuffed Crab 120
APPETIZERS
  Appetizers, Reheating frozen 14
  Almond Chicken Dip 23
  Avocado Sausage Squares 24
  Biscuit Bits 26
  Caramel Corn 20
  Chile con Queso 24
  Chili Pecan Log 26
  Curry Dip 23
  Hot Crabmeat Elegante 28
  Kabobs, Kaola Pineapple-Ham 27
  Manitaria Parayemista 22
  Mexican Pizza 25
  Mushrooms Vermouth 22
  Nutty Popcorn Balls 21
  Parmesan Popcorn 20
  Popcorn 19
  Sausage Con Queso 24
  Shrimp Butter 28
  Snowy Spiced Nuts 18
  Swedish Sausage Dip 23
  Texas Trash 21
APPLES
  Apple Crisp Pie 193
  Apple Nut Dressing 156
  Candied Apple Rings 159
  Cranapple Acorn Squash 149
  Healthy Streuseled Apples 158
  Red Candied Apples 159
Applesauce, Delicious 213
Apricarrots 135
Artichoke Halves with Dill Butter 124
Artichokes, Creamy Basil Dip for 125
Asparagus 125
Asparagus Turkey Spaghetti 113
Avgolemono Soup 38
Avocado Dressing 47
Avocado Sausage Squares 24
BABY FOOD, 211-213
  Applesauce, Delicious 213
  Baby Bottles 212
  Baby Food, Homemade 212
  Meats, Baby's First 213
  Peas, Pretty 213
  Squash, Acorn 213
BACON
  Bacon, Microwaving 98-99
  Bacon, Microwaving Chart 98
  Bacon and Beef Succotash 93
  Bacon and Egg Burrito 51
  Bacon Cheese Doggies, Jeff's 78
  Bacon Cheese Muffins 176
  Bacon Dressing, Hot 47
  Impossible Quiche Lorraine 55
  Lubbock Bacon Burgers 71
  Sunshine Breakfast Sandwiches, Bev's 51
  Veggie Topper 125

Baked Beans 127, 129
Baked Grapefruit 161
Baked Potato Skins 145
Baloney Burgers 75
BANANAS
  Chris' Banana Pudding 169
  Frozen Chocolate Bananas 160
  Rum Bananas 160
BAR COOKIES
  Giant Cookie Pizza 181
  Carrot Brownies 182
  Chocolate Chip Cookie Bars 180
  Chocolate Mint Brownies 182
  Coconut Bar Cookies 181
  Cranberry Torte 179
  Granola Bars, 6-Minute 180
Basil Beans 126
Basket Cases 178
Bavarian Cabbage 133
BEANS
  Basil Beans 126
  Celery Green Limas 127
  Chili Baked Beans 127
  Dried Beans 128
  Italian Green Beans 126
  Old Fashioned Baked Beans 129
BEEF
  Beef Bourguignonne 96
  Burgundy Steak Strips 94
  Cowpunchers' Beef Stew 94
  Round Steak Roll-Ups 95
Betty's Bonanza 104
BEVERAGES
  Cocoa Mix, Homemade 30
  Cranberry Wassail 29
  Individual Hot Chocolate 30
  Sleepytime Swiss Mocha Mix 31
  Tea, Cozy Rosé 32
  Tea Mix, Low-Calorie Spiced 32
Big Bird Soup 39
BISCUITS
  Biscuit Bits 26
  Cinnamon Biscuit Ring 175
Black Russian Cake 190
Braised Celery-Mushroom Combo 137
BREAD
  Garlic Bread, One Minute 61
  Pumpkin Bread 178
  Bread Defrosting, Hamburger Buns 70
  Bread Defrosting, Hot Dog Buns 76
  Breakfast Entrees, Reheating frozen 14
BROCCOLI
  Broccoli and Carrot Medley 132
  Broccoli Buttons and Bows 131
  Broccoli Ham Au Gratin 105
  Lemon Broccoli Spears 132
  Toadstools and Flowers 131
BROWNIES
  Carrot Brownies 182
  Chocolate Mint Brownies 182

Fudge Cake Squares with Chocolate Bar Frosting 183
Buckeye Candy 204
Burger Logs 74
Burgundy Steak Strips 94
CABBAGE
  Bavarian Cabbage 133
  Faster Than a Speeding Bullet 99
  Smokehouse Cabbage 133
CAKES
  Black Russian Cake 190
  Coconut Gridiron Cake 187
  Dapper Rabbit Cake 185
  Lemon Poppyseed Cake 188
  Lemonade Cake 188
  Masquerade Cake 191
  Mexican Chocolate Cake 184
  Nutty Cranberry Cake 191
  Pineapple Downside-Up Cake 186
  Prune Danish Snack Cake 179
  Streusel Bundt Cake Mixes 190
  Tourta Athineiki 189
  Yule Log 192
Candied Apple Rings 159
Candied Apple Yams 147
CANDY
  Buckeye Candy 204
  Chocolate-Covered Easter Eggs 207
  Chocolate Melting Chart 204
  Mocha Brandy Candy 206
  Never-Fail Fudge 205
  Peanut Brittle 208
  Peanut Butter Candy 206
  Pecan Brittle 208
  Red Candied Apples 159
  Rim Rock Crunch 208
  Rocky Road Fudge, Easy 205
  White Chocolate Pretzels 207
Caramel Corn 20
CARROTS
  Apricarrots 135
  Broccoli and Carrot Medley 132
  Carrot Brownies 182
  Citrus Spice Carrots 134
  Golden Coin Salad 44
  Honey Gingered Carrots 134
  Mr. McGregor's Garden Special 135
Cashew Tuna Casseroles 121
CASSEROLES
  Casserole Freezing Directions 101
  Casserole Technique 15
  American Mah-Jong 68
  Asparagus Turkey Spaghetti 113
  Bacon and Beef Succotash 93
  Betty's Bonanza 104
  Cheddar Shrimp Casserole 118
  Chicken & Noodles Paprika 112
  Chili Noodle Dinner 89
  Crunchy Celery Casserole 137
  Gaucho Casserole 89
  Golden Potato Casserole 144

217

Hopscotch Hot Dog Casserole 79
Inside-Out "Manicotti" 63
Mexican Manicotti 66
Nacho Fiesta Casserole 68
Now-and-Later Ham Casseroles 101
Sandy's Ham Strata 103
Sausage Casseroles, Freezer 100
Sausage Ragout 100
Taco Casserole San Antone 90
Tex-Mex Tortilla Casserole 89
Tortilla Lasagne 91
Tuna Casseroles, Cashew 121
Tuna-Noodle Ring 121
CAULIFLOWER
   Indian Spiced Cauliflower 136
CELERY
   Braised Celery-Mushroom Combo 137
   Celery Green Limas 127
   Crunchy Celery Casserole 137
   Mr. McGregor's Garden Special 135
Chalupa Salad 45
CHARTS
   Bacon Microwaving Chart 98
   Chocolate Melting Chart 204
   Corn Microwaving Chart 138
   Fresh Vegetable Microwaving Chart 10
   Frozen Convenience Food Chart 14
   Grapefruit Microwaving Chart 161
   Homemade Cocoa Mix Chart 30
   Hot Dog Microwaving Chart 77
   Low-Calorie Spiced Tea Mix Chart 32
   Meat Microwaving Chart 10
   Power Level Chart 8
   Power Level Setting Chart 9
   Reheating, Best Power Levels 16
   Sleepytime Mocha Mix Chart 32
CHEESE
   American Mah-Jong 68
   Cathy's Cheddar Fondue 67
   Cheddar Shrimp Casserole 118
   Cheese Manicotti, Sam's 65
   Chile con Queso 24
   Chili Pecan Log 26
   Fondue, Veggie 151
   Nacho Fiesta Casserole 68
   Onion-Tomato Cheese Pie 52
   Pizza Fondue 67
   Roquefort Burger Topping 72
   Vegetable Soup, Swiss Cheese 35
CHEESECAKE
   Microwave Springform Pan 166
   Chimpanzee Cheesecake 167
   CiCi's Chocolate Amaretto Cheesecake 168
   Daiquiri Cheesecake 167
   Marble Cheesecake 166
   Sugarless Cheesecake 165
Cheesecake Ice Cream 172

Cheesy Cod, Easy 115
Cheesy Escalloped Onions 142
Cheesy Meat Loaf 85
Cheesy Spinach 148
CHICKEN
   Almond Chicken Dip 23
   Chicken, Microwave to Grill 81
   Chicken, Stewing 38
   Chicken & Noodles Paprika 112
   Chicken 'N Stuffin' 107
   Chicken Cashew 112
   Chicken Cutlets Mozzarella 110
   Chicken Hungarian 109
   Chicken Pie, Chilled 46
   Chicken-Rice Soup, Hearty 38
   Cornish Hens, Apple Stuffed 108
   Crispy Butternut Chicken 107
   Fried Chicken, Frozen 13
   Gingered Chicken 111
   Honeyed Chicken 109
   Macadamia Nut Chicken 111
   Supreme Rolls 110
   Teriyaki Chicken 108
Chihuahua Dogs 79
Chile con Queso 24
Chili Baked Beans 127
Chili From a Mix 90
Chili Noodle Dinner 88
Chili Pecan Log 26
Chimpanzee Cheesecake 167
Chinese Turkey 114
Chipper Fish 115
CHOCOLATE
   Chocolate Melting Chart 204
   Chocolate Chip Cookie Bars 180
   Chocolate Coconut Crust 198
   Chocolate Crumb Crust 201
   Chocolate Ice Cream, Judy's 171
   Chocolate Lovers' Pie 197
   Chocolate Mint Brownies 182
   Chocolate Peppermint Fondue 163
   Chocolate-Covered Easter Eggs 207
   CiCi's Chocolate Amaretto Cheesecake 168
   Cocoa Mix, Homemade 30
   Fudge Cake Squares with Chocolate Bar Frosting 183
   Fudge Pecan Pie 196
   Mexican Chocolate Cake 184
   Mexican Chocolate Frosting 184
   Never-Fail Fudge 205
   Rocky Road Fudge, Easy 205
   Sleepytime Swiss Mocha 31
Chris' Banana Pudding 169
Christmas Spice 29
CiCi's Chocolate Amaretto Cheesecake 168
Cinnamon Biscuit Ring 175
Citrus Spice Carrots 134
COCOA
   Cocoa Cadillac 30
   Cocoa Grasshopper 30
   Cocoa Mix, Homemade 30
   Individual Hot Chocolate 30
   Sleepytime Swiss Mocha Mix 31

Coconut Bar Cookies 181
Coconut Cream Pie 199
Coconut Gridiron Cake 187
Coffee Cake   Mincemeat Coffee Ring 175
Colache 139
Coney Island Hot Dogs 78
Cookies (see Bar Cookies)
CORN, 138-140
   Chart, Fresh Corn, 138
   Colache 139
   Corn and Mushrooms 140
   Corn on the Cob, Baked 138
   Creole Corn 140
   Polka Dot Corn Custard 139
Cornish Hens, Apple Stuffed 108
Covering Foods 11
Cowpunchers' Beef Stew 94
Crabmeat Elegante, Hot 28
CRANBERRIES
   Cranapple Acorn Squash 149
   Cranberry Jalapeño Jelly 209
   Cranberry Orange Relish 210
   Cranberry Torte 179
   Cranberry Waldorf Salad 43
   Cranberry Wassail 29
Creamy Basil Dip 125
Creole Corn 140
Crispy Butternut Chicken 107
Croutons, Zippy 47
Crumb Crust 201
Crumb-Coated Tenderloin 106
Crunchy Acorn Squash 149
Crunchy Celery Casserole 137
Cucumber Salad 42
Curried Fruit, Hot 162
Curry Dip 23
Curry Sauce 136
Daiquiri Cheesecake 167
Dapper Rabbit Cake 185
Dehydrating 215
Dilly Cheese Potatoes 146
DIPS
   Almond Chicken Dip 23
   Chile con Queso 24
   Creamy Basil Dip 125
   Curry Dip 23
   Mexican Pizza 25
   Swedish Sausage Dip 23
Dirty Rice Dressing 155
Donuts, Reheating frozen 14
DRIED BEANS 128-130
   Dried Beans, Speedy MW Soak 128
   Baked Beans, Old-Fashioned 129
   Limas 'N Ham 129
   New Orleans Red Beans 130
Drying Leaves 216
Duck Gumbo 40
Duck, Stewing 40
Egg Carton Gems 174
Eggplant, Unbelievably 141
EGGS
   Bacon and Egg Burrito 51
   Hamburger-Box Omelet 50
   Sunshine Breakfast Sandwiches, Bev's 51
Faster Than a Speeding Bullet 99
FISH (see also Shellfish)

Defrosting, 115
Cheesy Cod, Easy 115
Chipper Fish 115
Fish, Frozen 13, 14
Impossible Cheesy Tuna Pie 54
Red Snapper Ring-A-Round 116
Salmon Ring, Curried 117
Salmon Steaks, Wine-Poached 117
Spinach Stuffed Fillets 116
Tuna Casseroles, Cashew 121
Tuna Microquettes 122
Tuna Topper 122
Tuna-Noodle Ring 121
Flower Drying 215
Foil, Shielding with 183
FONDUE
Cathy's Cheddar Fondue 67
Chocolate Peppermint Fondue 163
Pizza Fondue 67
Veggie Fondue 151
Freezer to Microwave 58
French Onion Soup 35
French Style Peas 143
French Vanilla Ice Cream 171
Fresh Fruit Swirl 172
Fresh Vegetable Microwaving Chart 10
Fried Chicken, Frozen 13
Fried Rice 154
FROSTINGS
Frosting, Creamy Cheese 182
Mexican Chocolate Frosting 184
Simple Glaze for Bundt Cakes 189
Frozen Chocolate Bananas 160
FROZEN CONVENIENCE FOODS, 12-14
 Chart, Frozen Convenience Food, 14
Fish, Frozen 13
Fried Chicken, Frozen 13
Pizza, Frozen 13
T.V. Dinners, Frozen 13
Frozen Pumpkin Squares 202
Frozen Vegetables 12
FRUIT 158-164
Fruit Defrosting 158
Baked Grapefruit 161
Candied Apple Rings 159
Candied Apples, Red 159
Chris' Banana Pudding 169
Curried Fruit, Hot 162
Frozen Chocolate Bananas 160
Healthy Streuseled Apples 158
Nectarine Cobbler 161
Peach Betty 164
Pears with Almonds, Baked 162
Rum Bananas 160
Strawberries Floridian 163
Strawberry Fluff 164
Fudge Cake Squares with Chocolate Bar Frosting 183
Fudge Pecan Pie 196
Garden Patch Impossible Pie 55
Garlic Bread, One Minute 61
Gaucho Casserole 89
Gazpacho Salad, Molded 42
Gazpacho de Ann 34

Gelatin dissolving 41
German Potato Salad, Speedy 144
Giant Cookie Pizza 181
GIFTS
Caramel Corn 20
Chili Pecan Log 26
Christmas Spice 29
Cocoa Mix, Homemade 30
Cranberry Jalapeño Jelly 209
Croutons, Zippy 47
Gazpacho de Ann 34
Sleepytime Swiss Mocha Mix 31
Snowy Spiced Nuts 18
Strawberry Platter Jam 209
Strawberry Preserves, Super Easy 210
Tea Mix, Low-Calorie Spiced 32
Texas Caviar 210
Texas Trash 21
White Chocolate Pretzels 207
Ginger Snap Crust 201
Gingerbread, Quick Pumpkin 178
Gingered Chicken 111
Glass lids 11
Golden Coin Salad 44
Golden Potato Casserole 144
Granola Bars, 6-Minute 180
Grapefruit, Baked 161
Gravy, Homemade 95
GRILLING, Microwave to 80, 81
Grilled Chicken 81
Grilled Spareribs 81
Hamburgers 71
Kabobs, Fresh Veggie 152
GROUND BEEF RECIPES
14-Karat Meat Loaf 85
Ann Landers' Meat Loaf 84
Bacon and Beef Succotash 93
Burger Logs 74
Chalupa Salad 45
Cheesy Meat Loaf 85
Chili From a Mix 90
Chili Noodle Dinner 88
Gaucho Casserole 89
Hamburger Stew 93
Hamburgers Continental 73
Hamburgers Teriyaki 74
Hamburgers, Double-Up 73
Hamburgers 70, 71
Herbed Mini-Loaves 86
Kathleen's Brontoburgers 72
Lieutenant Kelly's Balboa Island Spaghetti 60
Lubbock Bacon Burgers 71
Meatball Minestrone 39
Mexican Manicotti 66
Mexican Pizza 25
Mini-Mex Meat Loaves 86
Pizza Joe 75
Pocket Tacos 92
Spaghetti Pie 61
Spaghetti Sauce, Easy Meatball 59
Stuffed Onion Cups 143
Stuffed Spaghetti Squash 88
Taco Casserole San Antone 90
Taco Grande Pie 56
Tex-Mex Tortilla Casserole 89
Tortilla Lasagne 91
Vegetable Meat Pie 92

HAM
Broccoli Ham Au Gratin 105
Ham 'N Cheese Patties 103
Ham Pineapple Stack-Ups 102
Ham Slices, Carolina 104
Hamburger-Box Omelet 50
Kabobs, Kaola Pineapple-Ham 27
Limas 'N Ham 129
Macaroni and Ham Salad 63
Now-and-Later Ham Casseroles 101
Sandy's Ham Strata 103
Squash Halves, Stuffed 102
Hamburger Stew 93
Hamburger-Box Omelet 50
HAMBURGERS 70-75
Hamburger Microwaving 70
Hamburger Reheating 71
Baloney Burgers 75
Burger Logs 74
Continental Hamburgers 73
Hamburgers Teriyaki 74
Hamburgers, Double-Up 73
Kathleen's Brontoburgers 72
Lubbock Bacon Burgers 71
Pizza Joe 75
Roquefort Burger Topping 72
Hawaiian Slaw 43
Healthy Streuseled Apples 158
Herb Drying 214
Herbed Mini-Loaves 86
Hollandaise Sauce, Food Processor 131
Honey Gingered Carrots 134
Honeyed Chicken 109
HOT DOGS 76-79
Hot Dog Microwaving 77
Hot Dog Defrosting 76
Bacon Cheese Doggies, Jeff's 78
Chihuahua Dogs 79
Coney Island Hot Dogs 78
Hopscotch Hot Dog Casserole 79
Oscar J. Reubens 77
ICE CREAM 170-172
Cheesecake Ice Cream 172
Chocolate Ice Cream, Judy's 171
French Vanilla Ice Cream 171
Fresh Fruit Swirl 172
Frozen Pumpkin Squares 202
Ice Cream, Homemade 170
Mud Pie 203
Impossible Cheesy Tuna Pie 54
Impossible Quiche Lorraine 55
Indian Spiced Cauliflower 136
Inside-Out "Manicotti" 63
Italian Green Beans 126
Kabobs, Fresh Veggie 152
Kabobs, Kaola Pineapple-Ham 27
Kahlua Chiffon Pie 199
Kahlua Cocoa 30
Kartoffelrösti 145
Kathleen's Brontoburgers 72
Korean Dressing 44
Korean Spinach Salad 44
Lamb Loaf 87
Leftovers 16

Lemon Broccoli Spears 132
Lemon Meringue Pie 197
Lemon Poppyseed Cake 188
Lemonade Cake 188
Lieutenant Kelly's Balboa Island Spaghetti 60
Limas 'N Ham 129
Liver Cutlets 97
Liver Marengo 97
Lubbock Bacon Burgers 71
Lucky Peas 142
Macadamia Nut Chicken 111
Macaroni and Ham Salad 63
Manicotti, Microwaving 64
Manitaria Parayemista 22
Marble Cheesecake 166
Mashed Potatoes, Speedy 144
Masquerade Cake 191
MEAT LOAF, 83-87
   Microwaving Meat Loaves 83
   14-Karat Meat Loaf 85
   Ann Landers' Meat Loaf 84
   Cheesy Meat Loaf 85
   Herbed Mini-Loaves 86
   Lamb Loaf 87
   Mini-Mex Meat Loaves 86
   Turkey Meat Loaf 87
Meat Microwaving Chart 10
Meatball Minestrone 39
Meats, Baby's First 213
MEXICAN FOOD
   Chalupa Salad 45
   Chile con Queso 24
   Chili From a Mix 90
   Gaucho Casserole 89
   Mexican Chocolate Cake 184
   Mexican Chocolate Frosting 184
   Mexican Manicotti 66
   Mexican Mocha 31
   Mexican Pizza 25
   Mini-Mex Meat Loaves 86
   Nacho Fiesta Casserole 68
   Pocket Tacos 92
   Taco Casserole San Antone 90
   Taco Grande Pie 56
   Tex-Mex Tortilla Casserole 89
   Tortilla Lasagne 91
Microwave Timing 10
Microwave to Grill 81
Mincemeat Coffee Ring 175
Mini-Mex Meat Loaves 86
Mocha Brandy Candy 206
Moonshine Pie 201
Mr. McGregor's Garden Special 135
Mud Pie 203
MUFFINS
   Muffins, Basic 176
   Muffins, Reheating frozen 14
   Bacon Cheese Muffins 176
   Egg Carton Gems 174
   Oatmeal Raisin Muffins 177
   Strawberry Muffins 177
MUSHROOMS
   Braised Celery-Mushroom Combo 137
   Cindy's Mushroom Sauce 82
   Corn and Mushrooms 140
   Manitaria Parayemista 22
   Mushroom Bordelaise Sauce 82

   Mushrooms Vermouth 22
   Toadstools and Flowers 131
   Zucchini Boats 150
Nacho Fiesta Casserole 68
Nectarine Cobbler 161
Never-Fail Fudge 205
New Orleans Red Beans 130
Now-and-Later Ham Casseroles 101
NUTS, 18
   Peanut Pie Crust 203
   Pecan Crust 203
   Snowy Spiced Nuts 18
   Williamsburg Peanut Soup 37
Nutty Cranberry Cake 191
Nutty Popcorn Balls 21
Oatmeal-Raisin Muffins 177
ONION
   Cheesy Escalloped Onions 142
   French Onion Soup 35
   Onion-Tomato Cheese Pie 52
   Stuffed Onion Cups 143
Orange Rice 154
Oscar J. Reubens 77
Oyster Dressing 156
Paella Pronto 120
Pancakes, Reheating frozen 14
Paper toweling 11
Parmesan Popcorn 20
PASTA
   Pasta, Basic Information 58
   Pasta, Microwaving Directions 58
   Asparagus Turkey Spaghetti 113
   Cheese Manicotti, Sam's 65
   Chicken & Noodles Paprika 112
   Chili Noodle Dinner 88
   Inside-Out "Manicotti" 63
   Lieutenant Kelly's Balboa Island Spaghetti 60
   Macaroni and Ham Salad 63
   Mexican Manicotti 66
   Microwaving Manicotti 64
   Pasta Primavera 59
   Peppers, Spaghetti-Stuffed 62
   Sausage Manicotti Florentine 65
   Spaghetti Pie 61
   Spaghetti Sauce, Easy Meatball 59
   Spinach Fettuccine 62
   Tuna-Noodle Ring 121
Pastries, Reheating frozen 14
Peach Betty 164
Peach Pie, Fresh 194
Peaches 'n Cream Torte 194
Peachy Mincemeat Pie 195
Peachy Pretzel Pie 198
Peanut Brittle 208
Peanut Butter Candy 206
Peanut Butter Pie 202
Peanut Pie Crust 203
Pears with Almonds, Baked 162
PEAS
   French Style Peas 143
   Lucky Peas 142
   Peas, Pretty (Baby food) 213
Pecan Brittle 208
Pecan Crust 203
Peppers, Spaghetti-Stuffed 62

Percentages of Power 8
PIE CRUST
   Chocolate Coconut Crust 198
   Chocolate Crumb Crust 201
   Crumb Crust 201
   Gingersnap Crumb Crust 201
   Peanut Pie Crust 203
   Pecan Crust 203
   Pie Shells, frozen 193
   Pretzel Crust 198
PIES
   Apple Crisp Pie 193
   Chocolate Lovers' Pie 197
   Coconut Cream Pie 199
   Fudge Pecan Pie 196
   Kahlua Chiffon Pie 199
   Lemon Meringue Pie 197
   Moonshine Pie 201
   Mud Pie 203
   Peach Pie, Fresh 194
   Peaches 'n Cream Torte 194
   Peachy Mincemeat Pie 195
   Peachy Pretzel Pie 198
   Peanut Butter Pie 202
   Pumpkin Cheese Pie 195
   Pumpkin Chiffon Delight 200
   Rum Chiffon Pie 199
   Sweet Potato Pecan Pie 196
Pineapple Downside-Up Cake 186
Pineapple-Poppy Seed Dressing 48
Pizza Fondue 67
Pizza Joe 75
Pizza, Reheating frozen 14
Plastic wrap 11
Pocket Tacos 92
Polka Dot Corn Custard 139
POPCORN, 19-21
   Caramel Corn 20
   Nutty Popcorn Balls 21
   Parmesan Popcorn 20
   Texas Trash 21
PORK
   Crumb-Coated Tenderloin 106
   Microwave Grilled Spareribs 81
   Stuffing Top Chops 106
POTATOES
   Baked Potato Skins 145
   Candied Apple Yams 147
   Dilly Cheese Potatoes 146
   German Potato Salad, Speedy 144
   Golden Potato Casserole 144
   Kartoffelrösti 145
   Mashed Potatoes, Speedy 144
   Potato Soup, Baked 36
   Sweet Potato Pots 146
   Sweet Potatoes Calypso 147
Poultry, 107-114
Power Level Setting Chart 9
Power Levels for Reheating, Best 16
Power Levels for food, best 8
Power Levels 8
PRESERVES
   Cranberry Jalapeño Jelly 209
   Cranberry Orange Relish 210
   Strawberry Platter Jam 209
   Strawberry Preserves, Super Easy 210

Texas Caviar 210
Pretzel Crust 198
Prune Danish Snack Cake 179
PUDDING
   Chris' Banana Pudding 169
   Rice Pudding, Calcutta 169
PUMPKIN
   Frozen Pumpkin Squares 202
   Pumpkin Bread 178
   Pumpkin Cheese Pie 195
   Pumpkin Chiffon Delight 200
   Pumpkin Gingerbread, Quick 178
   Pumpkin Soup, Cream of 36
QUICHE
   Garden Patch Impossible Pie 55
   Impossible Cheesy Tuna Pie 54
   Impossible Quiche Lorraine 55
   Onion-Tomato Cheese Pie 52
   Shrimp and Avocado Quiche 53
   Spinach Quiche 53
   Taco Grande Pie 56
QUICK BREADS
   Basket Cases 178
   Cinnamon Biscuit Ring 175
   Egg Carton Gems 174
   Mincemeat Coffee Ring 175
   Pumpkin Bread 178
   Pumpkin Gingerbread, Quick 178
Red Candied Apples 159
Red Snapper Ring-A-Round 116
Reheating 16
RICE 153-155
   Dirty Rice Dressing 155
   Fried Rice 154
   Instant Rice 153
   Orange Rice 154
   Rice Pudding, Calcutta 169
   Spanish Rice 153
Rim Rock Crunch 208
Rocky Road Fudge, Easy 205
Roquefort Burger Topping 72
Round Steak Roll-Ups 95
Roux, Brown 40, 82
Rum Bananas 160
Rum Chiffon Pie 199
SALAD DRESSINGS
   Avocado Dressing 47
   Bacon, Dressing Hot 47
   Korean Dressing 44
   Pineapple-Poppy Seed Dressing 48
   Sue's Cooked Salad Dressing 48
SALADS
   Chalupa Salad 45
   Chicken Pie, Chilled 46
   Cranberry Waldorf Salad 43
   Croutons, Zippy 47
   Cucumber Salad 42
   German Potato Salad, Speedy 144
   Golden Coin Salad 44
   Hawaiian Slaw 43
   Korean Spinach Salad 44
   Macaroni and Ham Salad 63
   Molded Gazpacho Salad 42
   Shrimply Salad 46
   Turkey Salad, Hot 45

White and Blueberry Salad 41
Salmon Ring, Curried 117
Salmon Steaks, Wine-Poached 117
Sandy's Ham Strata 103
SAUCES
   Cindy's Mushroom Sauce 82
   Curry Sauce 136
   Gravy, Homemade 95
   Hollandaise Sauce, Food Processor 131
   Mushroom Bordelaise Sauce 82
   Vinaigrette Sauce 113
SAUSAGE
   Avocado Sausage Squares 24
   Faster Than a Speeding Bullet 99
   Sausage Casseroles, Freezer 100
   Sausage con Queso 24
   Sausage Manicotti Florentine 65
   Sausage Ragout 100
   Swedish Sausage Dip 23
Scallops Bonaventure 119
SHELLFISH
   Annapolis Stuffed Crab 120
   Cheddar Shrimp Casserole 118
   Hot Crabmeat Elegante 28
   Paella Pronto 120
   Scallops Bonaventure 119
   Shrimp and Avocado Quiche 53
   Shrimp Butter 28
   Shrimp Sahib 119
   Shrimp with Cashews 118
   Shrimply Salad 46
Shielding with Foil 183
Simple Glaze for Bundt Cakes 189
Sleepytime Swiss Mocha Mix 31
Smokehouse Cabbage 133
Snowy Spiced Nuts 18
SOUPS
   Avgolemono Soup 38
   Big Bird Soup 39
   Chicken-Rice Soup, Hearty 38
   Cold Spinach Soup 34
   Duck Gumbo 40
   French Onion Soup 35
   Gazpacho de Ann 34
   Meatball Minestrone 39
   Potato Soup, Baked 36
   Pumpkin Soup, Cream of 36
   Vegetable Soup, Swiss Cheese 35
   Williamsburg Peanut Soup 37
Spaghetti Pie 61
Spaghetti Sauce, Easy Meatball 59
Spanish Rice 153
SPINACH
   Cheesy Spinach 148
   Korean Spinach Salad 44
   Sausage Manicotti Florentine 65
   Spinach Coriander 148
   Spinach Fettuccine 62
   Spinach Quiche 53
   Spinach Soup, Cold 34
   Spinach Stuffed Fillets 116
SQUASH
   Colache 139
   Cranapple Acorn Squash 149

Crunchy Acorn Squash 149
Squash, Acorn (Baby food) 213
Squash Halves, Stuffed 102
Stoplight Special 150
Stuffed Spaghetti Squash 88
Zucchini Boats 150
Zucchini Tomatoes 151
Stoplight Special 150
Strawberries Floridian 163
Strawberry Butter 177
Strawberry Fluff 164
Strawberry Muffins 177
Strawberry Platter Jam 209
Strawberry Preserves, Super Easy 210
Streusel Bundt Cake Mixes 190
Stuffed Onion Cups 143
Stuffed Spaghetti Squash 88
STUFFINGS
   Apple Nut Dressing 156
   Cornish Hens, Apple Stuffed 108
   Dirty Rice Dressing 155
   Oyster Dressing 156
   Stuffing Top Chops 106
   Stuffing, Chicken 'N 107
Sue's Cooked Salad Dressing 48
SUGARLESS DESSERTS
   Peach Betty 164
   Strawberry Fluff 164
   Sugarless Cheesecake 165
Sunshine Breakfast Sandwiches, Bev's 51
Supreme Rolls 110
Swedish Sausage Dip 23
Sweet Potato Pecan Pie 196
Sweet Potato Pots 146
Sweet Potatoes Calypso 147
T.V. Dinners, Reheating frozen 14
Taco Casserole San Antone 90
Taco Grande Pie 56
Tea, Cozy Rosé 32
Tea Mix, Low-Calorie Spiced 32
Teriyaki Chicken 108
Tex-Mex Tortilla Casserole 89
Texas Caviar 210
Texas Trash 21
Toadstools and Flowers 131
TOMATOES
   Stoplight Special 150
   Zucchini Tomatoes 151
Tortilla Lasagne 91
Tourta Athineiki 189
Tuna Casseroles, Cashew 121
Tuna Microquettes 122
Tuna Topper 122
Tuna-Noodle Ring 121
TURKEY
   Asparagus Turkey Spaghetti 113
   Big Bird Soup 39
   Chinese Turkey 114
   Turkey Hurry Curry 114
   Turkey Meat Loaf 87
   Turkey Salad, Hot 45
   Turkey Tenderloin Vinaigrette 113
Unbelievably Eggplant 141
Veal Bellevoir from Zurich 96

Vegetable Meat Pie 92
Vegetable Soup, Swiss Cheese 35
VEGETABLES
  Fresh Vegetable Microwaving Chart 10
  Fondue, Veggie 151
  Kabobs, Fresh Veggie 152
  Pasta Primavera 59

Vegetable Meat Pie 92
Veggie Potpourri 152
Vegetables, Microwaving frozen 14
Veggie Potpourri 152
Veggie Topper 125
Vinaigrette Sauce 113

Wax paper 11
White Chocolate Pretzels 207
White and Blueberry Salad 41
Williamsburg Peanut Soup 37
Yule Log 192
Zucchini Boats 150
Zucchini Tomatoes 151

MicroScope
P.O. Box 79762
Houston, Texas 77279

Please send _____ copies of **Micro Quick** @ $10.95 each	$ _____
Please send _____ copies of **Microwave Know-How** @ $10.95 each	$ _____
Add $1.50 per book for postage and handling	$ _____
(Texas residents add $.55 per book for sales tax.)	$ _____
TOTAL ENCLOSED	$ _____

Make checks payable to **MicroScope.**

PLEASE MAIL TO:

NAME _____

ADDRESS _____

CITY _____ STATE _____ ZIP _____

- -

MicroScope
P.O. Box 79762
Houston, Texas 77279

Please send _____ copies of **Micro Quick** @ $10.95 each	$ _____
Please send _____ copies of **Microwave Know-How** @ $10.95 each	$ _____
Add $1.50 per book for postage and handling	$ _____
(Texas residents add $.55 per book for sales tax.)	$ _____
TOTAL ENCLOSED	$ _____

Make checks payable to **MicroScope.**

PLEASE MAIL TO:

NAME _____

ADDRESS _____

CITY _____ STATE _____ ZIP _____

- -

MicroScope
P.O. Box 79762
Houston, Texas 77279

Please send _____ copies of **Micro Quick** @ $10.95 each	$ _____
Please send _____ copies of **Microwave Know-How** @ $10.95 each	$ _____
Add $1.50 per book for postage and handling	$ _____
(Texas residents add $.55 per book for sales tax.)	$ _____
TOTAL ENCLOSED	$ _____

Make checks payable to **MicroScope.**

PLEASE MAIL TO:

NAME _____

ADDRESS _____

CITY _____ STATE _____ ZIP _____